Speaking Science to Power

Speaking Science to Power

Responsible Researchers and Policymaking

Edited by
Rachel A. Epstein
and
Oliver Kaplan

OXFORD
UNIVERSITY PRESS

OXFORD
UNIVERSITY PRESS

Great Clarendon Street, Oxford, OX2 6DP,
United Kingdom

Oxford University Press is a department of the University of Oxford.
It furthers the University's objective of excellence in research, scholarship,
and education by publishing worldwide. Oxford is a registered trade mark of
Oxford University Press in the UK and in certain other countries

Published in the United States of America by Oxford University Press
198 Madison Avenue, New York, NY 10016, United States of America

British Library Cataloguing in Publication Data

Data available

Library of Congress Control Number: 2024937043

ISBN 9780198875185
ISBN 9780198877134 (pbk.)

DOI: 10.1093/oso/9780198875185.001.0001

Printed and bound by
CPI Group (UK) Ltd, Croydon, CR0 4YY

Preface and Acknowledgments

When researchers are *Speaking Science to Power*, they have the opportunity and privilege to contribute to improving public policies. They may even generate knowledge that contributes to the quality of democracy, aids in conflict resolution, or advances more equitable economic policy, among other issues. But they also have a responsibility to guard against misunderstandings, misuse, misconduct, and other potential harms—whether originating with themselves or with those in positions of power who would apply their ideas.

Scientific research is a process involving agnosticism, logic, validation, peer input, correction, and at times the abandonment of notions perhaps once held dear. And most of all, it involves the humility of not knowing. Scholarly research may be similar to journalism or intelligence analysis in its ability to inform the public and decision-makers. But to serve as an arbiter of knowledge, science depends on criticism and transparency, as well as an appreciation of the ways in which ideology and politics may shape research. Scientists must be mindful that their projects and results may be suffused with politics, even if unintended. And while it may seem unlikely that the application of science could cause trouble, the problems in the science-to-policy interface are all too common, precisely because of the sweeping societal changes—and related interests and incentives—that the application of science can bring. Science is thus also power, and it should be used responsibly.

The publication of this volume confronts these issues and represents the culmination of years of collaboration. It began with a Responsible Engagement grant from the Carnegie Corporation of New York in 2018, spearheaded by Cullen Hendrix, in which several colleagues at the Josef Korbel School of International Studies at the University of Denver served as coinvestigators. At the first workshop it was striking how many distinct "responsible engagement dilemmas" we had all encountered. Everyone had a story to tell. Participants discussed having had their work pulled from policy meetings at the last minute because the findings did not fit comfortably with the larger agenda. They confessed uncertainty about how to communicate ambiguous or conflicting results with policymakers for fear they might gravitate toward familiar solutions rather than more tailored ones. And they shared insights about just how common stridency and overgeneralization were in fields as diverse as economics, environmental science, and international security.

Some of us had long, varied, and direct experiences with policy engagement, while others had written about policymaking and distributional struggles in our academic research. Yet there was broad agreement that there was little guidance in our fields for researchers coming up the academic ranks about how to navigate the complex world of policy engagement. What should one do if a policymaker cherry-picks findings, seeks out academic input only after a policy decision has been made, or plans to use research to increase the effectiveness or lethality of military intervention? How can academics understand at what point in the complex chain of policymaking their research might be inserted, and to what end? Under what circumstances would PhDs—who are trained in the value of circumspection, methodological rigor, specifying study limits, and alternative explanations—make forceful claims on the basis of partial or even nonexistent evidence? And is it possible to plan for the risk that the results one is using to advise policymakers might be incorrect?

None of us had received formal training in responsible policy engagement, but we had all ended up interested in contributing to policy through our research. Our motivation in bringing the case studies together was to both provide guidance to others (versions of our younger selves!) and to initiate a broader conversation about the ethics of policy engagement, particularly for social scientists, but really for any scientist involved in public-oriented projects. This resulting compilation became an introspective conversation about policy and process rather than the scientific probing of hypotheses. As such, completing this project required the support, input, and patience of many.

We would like to begin by thanking Cullen Hendrix, with the essential help of Jill Schmieder Hereau, for writing the Responsible Engagement grant proposal. Cullen had been grappling with the ethics of policy engagement in his own research for some time, and his vision and instincts kick-started our process of dialogue on these challenging and personal reflections. Our collegial conversations would also not have been possible without the Carnegie Corporation of New York and the steadfast commitment of Stephen del Rosso to the ideas of bridging the gap between researchers and policymakers and believing in our mission to infuse that interaction with greater ethical awareness. We are extremely grateful to del Rosso and his colleagues Aaron Stanley and Mwangi Thuita for their support of our collaboration and the ways in which their input sharpened our thinking.

The book also benefitted enormously from colleagues at the Korbel School and beyond. Deborah Avant provided feedback on many parts of the project and has herself been an important innovator in the field of responsible engagement, reflecting on how our disciplines too often confine our thinking about political possibilities rather than introducing new options or ways of thinking. Naazneen Barma was a source of wisdom throughout this book's evolution. As one of the founders of the longstanding Bridging the Gap program, also supported by Carnegie, she shared insights about academic communication and the difficulties of imparting nuance in policy contexts. Korbel School dean Fritz Mayer was also supportive and generous with his time and input.

We are also grateful to several other colleagues who were instrumental as participants in project meetings, as discussants of the project at conferences, or who provided additional suggestions and insights: William Akoto, Regina Bateson, Christian Caryl, Alexander Downes, Brent Durbin, Danielle Gilbert, Jim Goldgeier, Julia Macdonald, Dan Mahanty, Bridget Moix, Brian O'Neill, and Ryan Powers. We are also indebted to several Korbel School students and staff who provided valuable assistance toward the production of this volume. We thank Sage Brewer, Gergana Kostadinova Law, Grace Naegelen, Regan Parish, Taubie Sanders, and Cassis Tingley. We also thank our editor, Dominic Byatt, at Oxford University Press for believing in this project, as well as Karen Bunn.

Finally, as editors, we thank the academic contributors to this project: Jay Lyall, George DeMartino, Tricia Olsen, and Tim Sisk. They took the risk alongside us of exposing their policy-bridging insecurities—and even, at times, failures—so that others could learn from them. We are also particularly grateful to our policymaker contributors, César Caballero Reinoso and Leanne Erdberg Steadman, who provide a view of the mysterious policymaker side of the "gap" and who were also willing to briefly cross over to the academic side by joining this volume. We also hold a special place in our hearts for Cristina Serna of the Peasant Workers Association of the Carare River (ATCC), who collaborated ("partnered") with Oliver Kaplan as part of a novel policy engagement process and who provided additional valuable advice to inform our thinking.

Contents

List of Figures

List of Tables

Contributor Biographies

César Caballero is the director of the Colombian survey company Cifras y Conceptos, SA, and was formerly the director of the Colombian census bureau (DANE). He previously served as an advisor to the Mayor of Bogotá and the Director of National Planning. He was also Dean of the Faculty of Social and Business Sciences of the Autonomous University of Manizales. He has a PhD in Human and Social Sciences from the Universidad Javeriana and a MA in Latin American Studies from the University of Oxford.

George F. DeMartino is Professor of Economics in the Josef Korbel School of International Studies at the University of Denver. His work on the ethical entailments of economic practice include the books *The Economist's Oath* (Oxford University Press), the coedited *Oxford Handbook of Professional Economic Ethics*, and, most recently, *The Tragic Science: How Economists Cause Harm (Even as They Aspire to do Good)* (University of Chicago Press).

Rachel A. Epstein is Professor of International Relations at the Josef Korbel School of International Studies at the University of Denver. She is the author of *In Pursuit of Liberalism: International Institutions in Postcommunist Europe* (Johns Hopkins University Press) and *Banking on Markets: Bank-State Ties in Europe and Beyond* (Oxford University Press), which received the Ed A. Hewett award in international political economy.

Leanne Erdberg Steadman is the senior director of Strategy and Impact at the Open Society Foundations. She was previously a director at the US Institute of Peace, focusing on violent extremism and peace processes. She served on the National Security Council staff at the White House in the Obama administration as a senior advisor on homeland security and director of African Affairs. She also held positions at the US Department of State and the Department of Homeland Security. Steadman holds a JD in international law from Boston University.

Oliver Kaplan is Associate Professor at the Josef Korbel School of International Studies at the University of Denver. He is the author *Resisting War: How Communities Protect Themselves* (Cambridge University Press), which examines how civilian communities organize to protect themselves from wartime violence. Kaplan was a Jennings Randolph Senior Fellow at the US Institute of

Peace and previously a postdoctoral Research Associate at Princeton University and at Stanford University. Kaplan received his PhD in political science from Stanford University.

Jason Lyall is the inaugural James Wright Chair of Transnational Studies and Associate Professor of Government at Dartmouth College, where he also directs the Political Violence FieldLab. He is the author of the award-winning *Divided Armies: Inequality and Battlefield Performance in Modern War* (Princeton University Press). His research examines how inequality affects battlefield performance in modern war, the effectiveness of humanitarian aid in conflict settings, and fieldwork ethics. He has conducted fieldwork in Afghanistan and Russia.

Tricia D. Olsen is Professor and Stassen Chair of World Peace at the University of Minnesota's Humphrey School of Public Affairs and Department of Political Science. She is also a faculty affiliate of the University of Denver's Sié Chéou-Kang Center for International Security and Diplomacy. Dr. Olsen's book, *Seeking Justice: Access to Remedy for Corporate Human Rights Abuse* (Cambridge University Press), provides an in-depth exploration of victims' access to remedy mechanisms for corporate human rights abuses.

Timothy D. Sisk is Professor of International and Comparative Politics at the Josef Korbel School of International Studies at the University of Denver. His research, teaching and, policy-oriented work focuses on armed conflict and political violence, together with understanding and evaluation of processes of conflict prevention, management, and peacebuilding in fragile and postwar contexts. He teaches graduate and undergraduate courses on comparative politics, politics of deeply divided societies, elections and conflict dynamics, and sport and international politics.

1

Responsible Policy Engagement

A New Imperative for Social Scientists

Rachel A. Epstein and Oliver Kaplan

> Your scientists were so preoccupied with whether or not they could, they didn't stop to think if they should.
>
> **Ian Malcolm, *Jurassic Park***

The year was 1945.[1] Scientists working on the Manhattan Project were developing the atomic bomb and hoping that a demonstration of its destructive power would finally bring World War II to an end. Director of the project, physicist J. Robert Oppenheimer, worried that such powerful weapons could spell doom for humanity. His role in the project would weigh on his conscience for the rest of his life. At one point he told President Harry Truman "I have blood on my hands." His colleague, Joseph Rotblat, left the project once he realized its ultimate objectives. Rotblat would subsequently dedicate himself to nonproliferation, believing that "[it is] the technician, not the commander in the field, who is at the heart of the arms race" (Iskander 2022).

Developing the atomic bomb is undoubtedly among the most consequential examples of the ethical perils of science (Schweber 2000). But it also embodies the choices and trade-offs researchers often face when deciding whether and how to engage in policy processes, including when formulating a research question at the inception of any project. Oppenheimer and Rotblat did not have a roadmap or guidebook on what research to conduct, let alone whether to share it or how. In their respective decisions of voice (Oppenheimer) and exit (Rotblat), questions around the harms of developing a deadly technology and using it against civilian populations

[1] We would like to thank Deborah Avant, Naazneen Barma, Regina Bateson, Christian Caryl, Alexander Downes, Danielle Gilbert, Cullen Hendrix, Jason Lyall, Fritz Mayer, and Tricia Olsen for their comments on earlier drafts of this introduction. All errors are our own.

Rachel A. Epstein and Oliver Kaplan, *Responsible Policy Engagement.* In: *Speaking Science to Power.* Edited by: Rachel A. Epstein and Oliver Kaplan, Oxford University Press. © Rachel A. Epstein and Oliver Kaplan (2024).
DOI: 10.1093/oso/9780198875185.003.0001

arose, as did concerns about the possible costs of a land invasion in Asia, the prolongation of conflict, and the risk that competitors would deploy the technology first. Fraught experiences like this one, along with some scientists' involvement with the US military during the Vietnam War, paved the way for the founding of guard-rail organizations such as the Union of Concerned Scientists, which advocate for the cautious and considered use of science in the public sphere.[2]

Despite some notable initiatives to encourage the responsible use of science, vast potential for ethical lapses in the application of research to policy remains. Academics do not have to be involved in building "the bomb" to encounter ethical dilemmas in policy engagement. Commonplace questions include whether and under what conditions advising activists may inadvertently contribute to their vulnerability, or if, when advising militaries, we unwittingly increase the lethality of their tactics. Furthermore, by drawing attention to falsehoods that crop up in public debates, do we improve the informational environment or, rather, amplify misinformation? Ethics-laden choices do not just arise for the "hard sciences," but are also intrinsic to social science advising in policy processes, where the misuse of findings by policymakers and other forms of overreach are also concerns. Despite this reality, surprisingly there is no shared discourse on how to manage research-to-policy interactions and achieve better outcomes.

Social scientists are often drawn to their subjects of study because they want to understand the reasons behind perplexing events and trends, both good and bad. What are the causes of war, economic growth, or discrimination? What are the best remedies for climate change, human rights violations, or global pandemics? Following from this curiosity, and given the urgency of contemporary issues, it seems logical for researchers to share their expertise and inform policy processes to enhance human welfare. Such interaction between academics and practitioners is often assumed to be an unalloyed benefit. Indeed, under the right circumstances, scholarly engagement in evaluating, debating, and formulating public policies can advance our knowledge and improve real-world outcomes. Examples include increased human longevity, recent trends toward reduced extreme poverty, as well as decreases in violent conflict (see, variously, World Bank 2016, 2020; Roser 2018; Center for Systemic Peace 2020; Fisher 2022).

[2] "Our [scientific] community has the additional responsibility to educate the public, to evaluate the long-term social consequences of its endeavor, and to provide guidance in the formation of relevant public policy. This is a role it has largely failed to fulfill and it can only do so if it enters the political arena ... even though the technological revolution has greatly benefited mankind, it has also released destructive forces that our society has failed to control." UCS. 1969. "Founding Document: Beyond March 4," https://www.ucsusa.org/about/history/founding-document-beyond-march-4.

In complex policymaking environments, however, scholars may struggle to make positive contributions. Academics routinely confront a series of "engagement dilemmas" as they interact with policy and policymakers that they may not have anticipated ahead of time. Collectively, the contributors here argue that it is rare to encounter "ideal" circumstances for applying scholarly knowledge to policy problems. One must navigate institutions, competing interests, and power rivalries. And these difficulties are on top of the prior challenge of reaching scientific consensus or convergence on values. Not only is it difficult to apply sound research, but some episodes of academic engagement with policymakers have actually *done harm*. It is for this reason we foreground the observation by the Ian Malcolm character (portrayed by Jeffrey Goldblum) from the film, *Jurassic Park*, that the invitation to engage in policy processes should not compel engagement. Rather, researchers should first anticipate the possible perils of engagement, in addition to the benefits.

Our aim with this book is to improve the transmission of academic expertise to policy processes by encouraging ethical reflexivity in research, teaching, and policy advising. Termed "responsible engagement," the approach of the authors begins from the premise that constructive policy engagement requires attention to the ethical dilemmas that can arise as researchers either consult with or receive support from policymakers, or engage in policy processes directly. Ignoring these dilemmas can lead to practical and ethical problems that compromise good policymaking or lead to unintended consequences. Examples include risks that academics and policymakers alike may overgeneralize their findings to gain influence or win an argument, that political commitments remain unacknowledged within particular methodological and theoretical choices, or that power dynamics between investigators and research subjects marginalize some key stakeholders. On a more personal level, members of our research team have also encountered social and political pressure against presenting unpopular findings, worked within disciplines that accept or even encourage certain kinds of dissembling, and wrestled with the prospect that carefully conducted research could be used for unethical ends.

As the contributors to this volume attest, researchers have relatively little control over their engagement contexts, even if they are the ones producing or conveying the data (though control over data can be one source of researcher leverage). They may also not have a clear view of the larger policymaking processes in which they operate or may be unaware of the specific point at which a research intervention occurs in a complex chain of decision-making (see Erdberg Steadman, Chapter 9, this volume). Scholars can, however, anticipate the potential ethical dilemmas ahead of time. They can also think through possible appropriate responses, including in dialogue with stakeholders, peers, and colleagues. Our encouragement of readers to anticipate the possible perils of engagement stems from our experience: if contemplation of the ethical dilemmas begins only when one is already enmeshed in the policymaking mix, it may be too late to react effectively. By reasoning through one's preferred responses ahead of time, in the heat of the moment a researcher will be better able to avoid responding to social and other pressures in ways they may later regret.

Some of the chapters offer specific suggestions for researchers confronting potential dilemmas. One recommendation is to publish one's ethical redlines before conducting research in a conflict setting (Lyall, Chapter 2, this volume); another is making one's political commitments explicit as a partial response to overgeneralization (Epstein, Chapter 7, this volume). When no tailored suggestions are evident, we at least seek to share the ethical and practical dilemmas that we have encountered to assist others in anticipating such challenges and formulating their own responses in advance. We hope that by presenting a range of scenarios, scholars can better inoculate themselves against social pressure and avoid enabling poor policy choices.

Given our own training and experiences, we are aware that many PhD candidates and early career academics in economics, International Relations, and political science receive little or no formal mentorship on general professional ethics or, more specifically, in "responsible engagement" with policy processes. Some of us were even actively discouraged from conducting applied research. In many of our programs, theory, method, and objectivity were prized, which, while beneficial in some ways, tended to elide the power dynamics underpinning research design options and coding choices (Stone 2018; Zvobgo and Loken 2020; Gani and Marshall 2022), conveying a false sense of ethical neutrality. We realize that scholars working within feminist, Marxist, poststructuralist, or other epistemologies may not struggle to the same degree with striving for objectivity on the one hand while privileging certain values on the other. For various strains of positivists, however, there are tensions between values and retaining objectivity (Kirshner 2022: 40). We do not collectively advocate for any particular approach to conducting social science research—or to causality, description, theory, or interpretation—but instead simply encourage greater reflexivity, introspection, and critical review of one's previous work with an eye toward continuous improvement, regardless of one's scholarly identity.

Ironically, although many academics who produce policy-relevant research today were not trained in responsible engagement, there is increasing pressure from many academic departments for faculty to be policy relevant. The contributors here agree that social science research could and should be used productively to improve outcomes, yet this volume stresses that it is also important to know when to walk away from an engagement that is not ethically sound. As Hendrix et al. (2023) and Barma and Goldgeier (2022) point out, there can be powerful market rewards for academics who engage. Enlarging one's own academic or public intellectual reputation can redound to departments and universities. But incentives for scholars to develop public personae in their areas of expertise makes it all the more important that they pursue open dialogue and receive feedback from peers about the ethical redlines they should not cross.

There has been a growing interest in research ethics focusing on researchers' duties and best practices during data collection, particularly in conflict and postconflict settings. But there is still little content on ethical policy engagement. Existing "Bridging the Gap" initiatives—upon which we rely but also build—have traditionally been oriented around the mechanics of engagement (George 1993). They cover such topics as how to write for nonacademic audiences, prepare effective policy briefs, and master digital media. We connect with these literatures while expanding the conversation to identify the crosscutting engagement dilemmas found in debates as diverse as those over central bank responses to financial crises, remedying corporate human rights abuses, or assessing the effects of NATO enlargement (referring to the North Atlantic Treaty Organization).

Like many social scientists, the contributors to this volume are motivated by scholarly puzzles as well as the desire to improve human welfare through research. To a person, our authors have found that engaging in policy processes productively is far more complex than their earliest interests in politics led them to anticipate. Attempting to make sense of the intersection of knowledge, power, and positionality in any given policy engagement can feel overwhelming, particularly on one's own. Our purpose here is threefold: to encourage researchers (as well as policymaking consumers of research) to grapple with that intersection, no matter how daunting; to ensure that no one is alone in deciding whether or how to engage; and to provide a resource to ponder possible engagement dilemmas early on, well before engagement begins. The case studies presented here are an invitation to readers to join in an ongoing conversation about the dilemmas and lessons we have encountered. We begin by identifying the engagement dilemmas and by situating our challenges within the larger literatures on policy engagement, science communication, and "Bridging the Gap."

Identifying Engagement Dilemmas

This book identifies six distinct engagement dilemmas that the contributors have confronted. The first two concern how researchers conduct themselves, often in the face of social and political pressure. The third and fourth dilemmas explore how researchers navigate seemingly irreconcilable imperatives as they engage with policy issues. The final two dilemmas address disciplinary tendencies that can compromise the quality of academic contributions to policy processes or undermine public confidence in expert analysis. In addition to the case studies, two reflections from policymakers help illustrate how these dilemmas have played out in real-world examples. The dilemmas, their thematic categories, the policy issues they involve, and the advice for addressing them are displayed in Table 1.1.

Table 1.1 Responsible policy engagement dilemmas

Thematic Category	Dilemma	Policy Issue	Advice
Social pressure	Seduction (Jason Lyall)	Military intervention	Know and publish your ethical redlines before conducting research
	Unpopular findings (Tricia Olsen)	Business and human rights abuse/remedies	Examine the position of policy interlocutors within the organization; understand the politics of the issue at stake
Competing imperatives	Ambiguous evidence or competing claims (Tim Sisk)	Electoral design for peacebuilding	Present contingencies of findings; advise policymakers of the advantages and pitfalls of generalization
	Positionality (Oliver Kaplan)	Civilian self-protection and community partnerships	Design research and dissemination for inclusion; weigh costs and benefits to research subjects
Affirming transparency norms/ challenging others	Prosocial lying (George DeMartino)	Economic crisis response	Respect autonomy of recipients of information; critically examine disciplinary norms; refrain from prosocial lying
	Overgeneralization (Rachel A. Epstein)	NATO enlargement	Recommit to research norms; be transparent about political commitments; critically review one's own work

The engagement dilemmas we identify here are rooted in key structural conditions of policy interactions that inevitably arise from social pressures, competing imperatives, and the neglect of disciplinary norms. Beginning with researcher conduct under social pressure, one of our contributors focuses on the problem of "seduction" (Jason Lyall; Chapter 2). Another highlights the problem of presenting unpopular findings to sponsors of the research (Tricia Olsen; Chapter 3). "Seduction" refers to researchers engaging in unethical practices or allowing their findings to be used for questionable ends as a result of the research context's normalization of values that undermine responsible engagement. In other words, researchers might succumb to unethical practices in the hopes of retaining access to a field setting or to maintain certain relationships. Seduction can come into play when researchers have not mapped out their ethical boundaries in advance, particularly if their access to information depends on embedding within agencies, militaries, or communities.

Two additional dilemmas also involve difficult-to-reconcile imperatives. First, when there are ambiguous or conflicting results from research, scholars may be challenged to help policymakers exercise their best judgment (Tim Sisk; Chapter 4). Second, the problem of positionality (Oliver Kaplan; Chapter 5) refers to conditions under which there are asymmetries of power and access in research and policymaking. Academics—especially those from the Global North—may partner with allies who do not enjoy the same degrees of power and privilege to inform policy processes. In such situations, the pressure from stakeholders is indirect, as researchers *should* feel pressure to consider their ethical obligations to those they study. With such power and resource disparities, scholars are advised to critically examine their positions and identities, which in some cases may militate against policy engagement, no matter how well intentioned.

The last two engagement dilemmas arise when social science enters in the heat of policy debates or policymaking, or, more generally, when researchers present their findings. The problem of "prosocial lying" (George DeMartino; Chapter 6) is closely linked to concerns about "overgeneralization" (Rachel A. Epstein; Chapter 7). Social scientists are trained to guard against both overgeneralization and deception—ensuring that their conclusions and recommendations are consistent with the scope of their cases—through careful causal specification, exploration of alternative explanations, and use of rigorous methods. These are the formal norms of social science. And yet the drive to wield influence too often leads researchers, even very esteemed ones, to *overstate* their cases, in violation of these norms. Prosocial lying may also be patronizing toward publics as researchers and policymakers alike conceal probable future realities, trying (in their minds), to curb panic or otherwise shape behavior in ways that they believe are beneficial. Both of these problems can lead to policy choices with weak evidentiary support.

Finally, in addition to exploring the six distinctive engagement dilemmas, two policymakers share their experiences and wisdom, both in conducting and consuming research. César Caballero Reinoso (Chapter 8) recounts the serious dangers he encountered as the data he developed in Colombia as a public official and later as a pollster became politicized in that polarized society. His experience embodies an extreme case of the "unpopular findings" dilemma and the risks of prosocial lying. It also illustrates how positionality can play out as some scholars and practitioners may confront much more direct and severe negative consequences than others. Leanne Erdberg Steadman's contribution (Chapter 9) reflects on what researchers should know about policy processes before they engage. It shows how complex most processes are and how critical the timing of researcher input is to impact. She also provides a reality check on how effective conscientious refusals to engage actually are, since if a researcher is called upon to provide input, it is likely that policymakers are already consuming that scholar's work. In-person engagement in those cases may simply allow the scholar to contextualize the findings.

When Facts Don't Always Speak for Themselves: Situating "Responsible Engagement"

Our volume charts a middle path between two different perspectives in the literature on evidence-based policy engagement. The first perspective suggests that scholars have a high degree of choice and autonomy in their mode of engagement, with facts or findings at the core of their positions (Pielke 2007). The second, opposing view maintains that scientific input—those very facts and findings—too often fails to gain traction in policy processes because facts *do not* speak for themselves and researchers do not control if or how their findings are absorbed into policymaking (Mair et al. 2019). Indeed, scholars of science communication have documented how identity, values, and emotion infuse human reasoning and affect the acceptance—or rejection—of facts (Lewandowsky et al. 2020; Scharfbillig et al. 2021). Our middle path acknowledges that a researcher's choice, autonomy, and adherence to facts may be unexpectedly upended precisely because their findings are not communicated effectively. By anticipating engagement dilemmas, we aim to help scholars and practitioners think through problems before they encounter them on the ground, thereby improving the conveyance of findings. But even as we acknowledge the importance of effective communication, we also realize that not all academics are equipped to become experts in science communication. However filtered, the "facts" as our contributors see them remain the coin of their research realm.

We outline the literature on academic policy engagement, focusing on international affairs and starting with the perspective that one can choose a well-defined role in policy processes (Pielke 2007). We also explain why that view is contested, since the science communication literature questions whether the highly informed, fact-infused researcher can effectively engage in policy processes where there is high political polarization. Ultimately, we conclude that there is a path between these two competing perspectives that most academics can take: becoming thoughtful, informed, and ethically reflective gap-bridging scholars.

Lots of Engagement, But Little Preparation

Over the last decade, International Relations scholars have debated whether their research has become increasingly irrelevant to the policy world and whether some kinds of social science (in economics, comparative politics, sociology, etc.) are better able to command the attention of policymakers. Avey et al. (2022) find that policymakers do routinely engage with academic research, though government officials in the areas of trade and development report greater reliance on scholarly work than those in national security. Although the "Ivory Tower/Beltway" divide is not as large as some have suggested, policy analysis—which policymakers find among the most useful outputs by researchers—has been in decline since 1980 (Avey et al. 2022: 16). There is therefore a gap between what scholars suspect is most useful for policymakers (policy analysis with policy prescriptions) and what academics choose to produce ("basic" research that advances debate within their disciplines).

At the same time, International Relations scholars demonstrate a strong appetite for relevance and engagement. A group of scholars recently assessed the extent of the policy impact of research across different topics of International Relations (Maliniak et al. 2020). In another recent study surveying scholars in international affairs, a large majority (70 percent) reported engaging in policy processes (Hendrix et al. 2023: 17). While only a very small minority reported that they themselves had ever adjusted their articulation of views to gain traction in policy processes (4.7 percent), more than a third of respondents (36.4 percent) expressed the fear that their peers routinely do so. This is problematic because it suggests that, in at least some instances, researchers provide distorted versions of their beliefs to policymakers to gain influence, regardless of what their research actually concludes. Doubts about researchers' conduct along these lines, including among academics, could also undermine the credibility of social science findings, as viewed by experts and the public. Moreover, it is possible that scholars underreport their inclinations to self-censor, or they may be unaware that they are distorting their views. Social pressure to conform in the reporting of results can be powerful.

The same study also found that the more advanced academics were in their careers, the more likely they were to value their own findings above a scholarly consensus when the two were in conflict (see Hendrix et al. 2023: 26–28). Given how widespread policy engagement turns out to be and how potentially ethically problematic it is to distort of one's views or neglect competing claims, we agree with these authors that the *lack* of attention to and training in responsible engagement is indeed "striking" (Hendrix et al. 2023: 26). This volume seeks to fill this gap in ethical awareness.

The Limits of Choice and Autonomy in Policy Engagement

Roger Pielke's 2007 book, *The Honest Broker*, is an important starting guide to policy engagement. He argues that explicitly acknowledging the precise role one seeks to play in advising or policymaking increases one's effectiveness and improves the quality of discourse (see also Olsen, Chapter 3, this volume). As contributors to this volume would attest, too many policy debates devolve into *ad hominem* attacks, enable the politicization of findings, or construct opposing sides that fail to address the substantive issue at stake. Pielke's book seeks to help academics avoid such pitfalls through awareness of their roles.

Pielke outlines four productive modes for researchers vis-à-vis the policy process. Given the nature of their work, *pure scientists* choose from the outset not to engage directly but rather to allow research to serve as a resource. *Science arbiters*, by contrast, answer policymakers' questions directly and thoroughly, relying on their knowledge of facts and without reference to values or their own preferences. *Issue advocates* also have good command of findings and debates, but use their expertise to limit the scope of plausible alternatives for policymakers by articulating why some options are better than others. Finally, *honest brokers* of policy options seek to expand or clarify the scope of policy alternatives available to policymakers. Because of scientific specialization, this approach often occurs through a committee or panel to allow experts in different areas to present arguments for each of the alternatives.

Pielke argues further that it is important for researchers to understand the nature of the issue at stake. As much as scholars might strive for objectivity, completely separating science from politics is nearly impossible (Jasanoff 1990; Pielke 2007), though some issues are clearly more politicized than others. Two key variables here include whether there is high or low scientific certainty around an issue and whether there is strong consensus on values. Are you, as a researcher, entering a debate that resembles the politics surrounding tornadoes characterized by scientific agreement and values consensus, or abortion politics, where there is neither? Pielke's central claim is that by specifying roles and understanding the nature of issues, researchers can optimize their contributions. Most crucially, they can (and should) avoid becoming "stealth issue advocates" who either knowingly or unwittingly hide values and preferences behind science, using their knowledge not to inform, but to manipulate (see Epstein, Chapter 7, and DeMartino, Chapter 6, this volume).

We agree with Pielke's insights and endorse the idea that bringing deep awareness of one's role and the issue at stake is essential. We also agree that researchers have power through their data and analysis vis-à-vis policymakers. But we carry Pielke's analysis forward and sometimes depart from it in three ways. First, as a group of social scientists, every debate in which we conduct research is closer to abortion politics than to tornado politics since our fields often feature little scientific consensus and values diverge. Indeed, in the last fifteen years, issues that we once thought resembled tornado politics have become increasingly contested and politicized—a global pandemic (COVID-19) serving as the most recent and extreme example.

Second, whereas Pielke's argument emphasizes the high degree of autonomy, efficacy, and choice on the part of researchers, the case studies here (along with the reflections by two policy experts) emphasize the reality of contingency, uncertainty, and the unexpected. Researchers can do as Pielke suggests and try to select a preferred role and understand the politics of an issue. However, we show that those steps alone cannot always prepare scholars for the social pressure at the heart of our dilemmas: pressure to minimize disruptive findings, provide evidence for ethically questionable ends, advise policymakers in the face of competing claims, or share findings in conflict-affected communities (see, respectively, Olsen, Lyall, Sisk, and Kaplan, Chapters 3, 2, 4, and 5, this volume).

Our analysis departs from Pielke's in a third way, as we view policy engagement dilemmas as affecting not only the quality of discourse and action, but also the welfare of researchers themselves. While Pielke is very focused on how researchers can impact policy, he pays less attention to the risks that researchers face if they engage and encounter problems, such as the misuse of their findings. For example, Pielke's coverage of the US invasion of Iraq in 2003 makes clear that select US policymakers had decided to attack before marshalling flawed evidence of Iraqi weapons programs as justification. As Pielke argues, this compromised the perceived integrity of research and intelligence.

We also recognize that whenever researchers are brought into policymaking solely to substantiate a decision that was already made, they become merely props rather than genuine sources of expertise. This is demoralizing for the scholar and undermines academic research more broadly. While we are concerned about the quality of discourse and policy and strive to improve both, we also focus on the ways in which risks for researchers in policy processes affect, and in some cases diminish, the engagement experience. The two problems are not separable. By presenting engagement dilemmas at the individual level, we aim to empower researchers by helping them anticipate the full range of contingencies, while also improving the quality of policymaking.

The Value of Science Communication

Another stream of literature related to policy engagement is science communication, which examines how audiences respond to scientific findings. Given increasing polarization in many societies, scholars in science communication are committed to better understanding the conditions under which evidence-based policy gains traction or is rejected by publics (e.g., Pechar et al. 2018; Mair et al. 2019; Diamond et al. 2020; Lewandowsky et al. 2020; Scharfbillig et al. 2021). Effective communication of evidence-based policy for the public good is a growing concern given the apparent appeal of false narratives and exclusionary politics (e.g., Kruse and Zelizer 2022). Recent advances in fields ranging from the humanities to the natural and social sciences, including but not limited to organizational behavior, physiology, and social and political psychology, have led meta-analyses to conclude, simply, that "facts do not speak for themselves" (Mair et al. 2019: 45). In practice, this has meant widening gaps between scientific near-consensus on some issues and highly polarized public opinion (Pechar et al. 2018; Diamond et al. 2020).

The science communication literature emphasizes several insights that have not typically been embedded in social science training. The first is that metaphor, narrative, and thoughtful framing are necessary to ensure that information is not only heard, but also understood (Mair et al. 2019: 45; see also Mayer 2014). The second is that investigators have often been lured into believing that there is a dichotomy between reason and emotion when, in fact, research shows that emotions are "as likely to enhance rationality as to subvert it" (Mair et al. 2019: 30; see also Epstein, Chapter 7, this volume). A third conclusion is that underlying values strongly inform one's interpretation of facts (Scharfbillig et al. 2021). Fourth, and more familiar to political scientists and economists, is that there is no such thing as a neutral frame for issues or research findings. This literature therefore argues that it "is not the side with the most or best facts that wins an argument, but the one that provides the most plausible scenario that feels intuitively reliable, communicated by a perceived credible source" (Mair et al. 2019: 47; see also Gheciu 2005; Epstein 2008; Pechar et al. 2018).

In some ways, the science communication literature questions Pielke's assertion that there can be "pure scientists" or that "honest brokers" or even "issue advocates" are sufficiently equipped to persuade policymakers or publics of a particular course of action. And, indeed, many of the contributors to this volume confronted the limitations of engagement outlined in the science communication literature. For example, in their chapters on competing claims and unpopular findings, Tim Sisk and Tricia Olsen have both encountered audiences with entrenched political beliefs rooted in values, identities, and material interests. Jason Lyall and George DeMartino consider the benefits of publicizing ethical redlines and the damaging consequences of prosocial lying, respectively, and find that competing values within their research contexts led to irresponsible engagement by agencies or peers.

The case studies in this volume illustrate the contributions of the science communication literature while also charting a middle path between focusing on the "facts" as our researchers view them and the most effective mode of conveyance. The meta-analyses profiled here advance one side of the argument, which is that, for researchers, "[t]oo strong a belief in the ability to neutrally frame and express information may in fact be counterproductive in that it makes the communicator less conscious of their own frames and worldviews" (Mair et al. 2019: 50). But by contrast, in at least some instances, publics will update their views through reasoning motivated by accuracy. New scientific information can change people's minds on an issue and move them closer to scientific consensus, irrespective of the source's identity or the subject's prior skepticism toward science in general

(Diamond et al. 2020). The contributors to this volume are more representative of typical policy-engaged social science scholars, who are not experts in science communication. Our case studies raise awareness of what science communication offers, while acknowledging that there is no single proven formula for advancing evidence-based policymaking.

Responsibly Closing the Gap between What We Know and Policy

This volume on "responsible engagement" is closely aligned with the "Bridging the Gap" program, two of the founders of which have recently reflected on how International Relations scholars should *not* engage with policy processes (Barma and Goldgeier 2022). Focusing on the importance of influence, interlocutors, integrity, and inclusivity, Barma and Goldgeier argue that effective engagement results from long-term efforts to establish close working relationships with broad and diverse networks of scholars and policymakers. When academic advising goes wrong, it could be attributed to scholarly blind spots or to policymaker instrumentality.

Barma and Goldgeier point out, for example, that at least some of the early scholarly work on democratic peace theory was refined in ways that should have severely limited the scope of anyone's claims about it, including its applicability to policy (2022: 1775–1776).[3] This did not stop members of the George W. Bush administration, however, from inferring the *wrong* lessons from the democratic peace literature and overstating the expected salutary effects of regime change in Iraq (similar to Pielke's analysis; 2007). Another example of suboptimal engagement between researchers and the policy world was the perhaps too-close relationship between some researchers of peacebuilding and the UN in the 1990s. With the benefit of hindsight, Barma and Goldgeier show that, in some cases, scholars appeared to have uncritically internalized the normative appropriateness of the UN's peacebuilding model. Researchers' reliance on UN data and access to develop their findings and seek policy relevance might have compromised critical interrogation of the UN's peacebuilding agenda, also to the detriment of sound policy.

We share many of Barma and Goldgeier's concerns. Overgeneralization or misrepresentation by both academics and policymakers can lead to unfounded expectations around a policy's possible effects (DeMartino, Chapter 6; Epstein, Chapter 7, this volume). The seduction of researchers

[3] Recent scholarship has more directly challenged some of democratic peace theory's strongest claims, even concerning "established" democracies (Rathbun, Parker, and Pomeroy 2024).

by the promise of access and attention is antithetical to the scholarly enterprise (Lyall, Chapter 2, this volume). Investigators are rarely in a position to advance policy that challenges the status quo, even when their findings are robust (Berry and Lake 2021; Olsen, Chapter 3, this volume). Power asymmetries result in the exclusion of crucial insights, which undermines policy effectiveness (Kaplan, Chapter 5, this volume). This book explores these issues, alongside personal accounts of what we did to try to mitigate "engagement harm," or what we wished we had known in advance.

Plan for the Book

The case studies and policymaker reflections in this volume each speak to a range of issues. Through paired comparisons, we group them into four, more focused categories. These include 1) researchers confronting social pressures within policy contexts (Lyall [Chapter 2] and Olsen [Chapter 3]); 2) the difficulties of accommodating seemingly irreconcilable imperatives within the engagement enterprise (Sisk [Chapter 4] and Kaplan [Chapter 5]); 3) reinforcing formal disciplinary norms (while challenging informal ones) to improve the quality of policy debate (DeMartino [Chapter 6] and Epstein [Chapter 7]); and 4) understanding policymaking complexity to better define the researcher's role (Caballero [Chapter 8] and Erdberg Steadman [Chapter 9]). By raising awareness of these issues, our aim is to improve the transmission of academic expertise to policy processes by encouraging ethical reflexivity among researchers. This is distinct from the search for scientific consensus or crafting effective narratives about which policies best serve the public good, which doubtless also contribute to productive engagement.

Coping with Social and Political Pressure

It is common to hear stories about colleagues whose research was suddenly taken off a policy meeting agenda because the findings were controversial, or who were reluctant to be as critical in face-to-face meetings with policymakers as they are in their written work. In other cases, peers have been approached to do research for well-funded and high-profile agencies, but whose end goals were unethical, or at least questionable, by the researchers' standards. Such situations can cause social discomfort. But they can also risk a researcher's access to funding, research subjects, and avenues for research dissemination. We are not suggesting one can always anticipate when social pressure will interfere with policy engagement, but being aware of the risk can help one think through appropriate responses in advance.

Jason Lyall's chapter, "Pre-register Your Ethical Redlines: Vulnerable Populations, Policy Engagement, and the Perils of E-Hacking," argues that social pressures are especially acute and also very high-stakes in conflict settings or in the midst of contentious politics. Not only is access difficult for researchers, but findings could adversely affect populations. Conflicts of interest abound, as parties to a conflict potentially view researchers as one vehicle to advance their side of an argument. Drawing on field work experience in Russia and Afghanistan, as well as on work with a range of government agencies and nongovernmental organizations, Lyall makes a detailed set of recommendations around pre-registering one's ethical redlines and publishing them in advance of any field research to guard against the normalization of unethical practices in research settings. Publishing a "pre-registration" document provides an ethical anchor and reference point, bringing accountability to the researcher and warding off dubious research and policy partners in the first place. It also forces the investigator to think through, in dialogue with peers and possibly also research subjects, the potential harms to populations that could inadvertently flow from any given project. "E-hacking" (ethics hacking) refers to ethical compromises that researchers at times feel pressure to make, while "preregistration" is one method of inoculating oneself against such pressure (i.e., avoiding "seduction").

Writing about a different kind of social pressure, Tricia Olsen's chapter is on "Sharing Unpopular Findings: When Academic Research Challenges Policy Efforts." Olsen recounts her experience of being approached by the United Nations Working Group (UNWG) on Business and Human Rights about creating a new dataset on corporations and their human rights records. The seeming synergies linking the UNWG's professed commitment to data-driven policymaking to Olsen's own research interests appeared ideal to her (and her coinvestigator) at the time. In retrospect, however, it became clear that the UNWG faced constraints that led them to disengage from findings that did not directly support the notion that the UN's Guiding Principles on Business and Human Rights shaped corporate behavior in ways that protected human rights. When that was *not* among the core conclusions, Olsen confronted a range of questions, including how a researcher shares unpopular findings with stakeholders when they conflict with policies that are already agreed upon and/or undergoing implementation. The question applies equally when researchers need the good will of organizations to ensure better policy outcomes or continued financial support of research. Olsen's chapter contextualizes the importance of probing the politics of an issue ahead of time, as well as understanding the placement of actors within their organizations. Without changing anything about the integrity of the research itself, Olsen provides invaluable guidance on strategies for engagement when incentives between scholars and policymakers are not aligned.

Reconciling Competing Imperatives

Reconciling competing imperatives in the research-to-policy interface can complicate responsible engagement for academics. One example concerns how scholars should conduct themselves within policy debates when the literature postulates conflicting claims about the best course of action. The dilemma for researchers is three-dimensional: the scholar 1) doubtless has their own findings, 2) should nevertheless provide policymakers with the array of plausible alternatives, but also 3) would ideally contribute to an informed path forward, even as risks are intrinsic to any policy choice given the highly contingent nature of many research findings. Competing imperatives also arise with respect to a researcher's positionality. Sometimes the competing imperatives are as fundamental as between the search for knowledge about a community and the perceived exploitation of a community that results from conducting research. Typically, however, the trade-offs are more subtle. Researchers are often in a position to empower the stakeholders who are the subjects of research. But not simultaneously marginalizing others is a challenge, as is avoiding inadvertently harming those whom research elevates. Engaging responsibly requires that scholars carefully consider whether they face competing imperatives and how best to navigate them.

Timothy Sisk's chapter, "Coping with Conflicting Claims: Ethnic Power Sharing for Sustainable Peace?" reflects on how to advise policymakers in the absence of scholarly consensus. The analysis explores whether ethnically based power sharing is advisable in postconflict and deeply divided societies. While there is evidence that such an approach can "stop the killing," many studies also point to the potential for cementing ethnic divisions rather than the effective cultivation of crosscutting cleavages. Sisk systematically reviews debates over how peace accords shape downstream institutional performance, the appeal and perils of presidentialism, trade-offs within different electoral systems, and the possible advantages and risks of power decentralization. Sisk concludes that while policymakers may be attracted to scholarly findings identifying the "best" institutions for cementing lasting peace, their risk aversion may nevertheless steer them toward suboptimal solutions to arrest violence in the short term. While competing imperatives may not be resolvable, Sisk argues that researchers can still engage responsibly by highlighting contingency and context in connection with every finding. Explaining to policymakers the extent to which particular methodological choices apply (or not) to any given policy setting can potentially improve the quality of institutional design.

Oliver Kaplan's chapter, "Partnered Policy Engagement: Positionality and the Ethics of Collaborating with Civil Society Organizations," explores how scholars can in some cases transcend competing imperatives. His own experience empowered—rather than marginalized—stakeholders who were the subjects of research. Through a collaboration among academics, local peacebuilders, and NGOs, Kaplan shared the successes and failures of community-based atrocities prevention strategies with other NGOs and government policymakers. He further reviews the ethical considerations involved in what he terms his process of "partnered engagement" with a local peacebuilder from Colombia, Cristina Serna. Kaplan and Serna co-presented their findings and experiences at in-person briefings in the US with NGOs and policymakers. The connections with the policymakers proved highly valuable for Serna and her community, allowing her to share lessons with a broader audience and gain support from new international actors. The participation of Serna's local voice and face in the dialogues also gave more weight to and humanized the available scholarly research findings. At the same time, the partnering was not without risks and possible negative effects. The process elevated some countries, communities, and individuals over others in terms of benefitting from the engagement opportunity, and the travel also posed opportunity costs and potential hardships. Kaplan concludes that for partnered engagements to be both successful and responsible, scholars should address the potential burdens on local actors as well as researcher positionality.

Affirming Transparency Norms While Challenging Other Norms

Some existing norms in social science research bolster the quality of the academic–policy interface, while others undermine it. Milton Friedman (1953) believed that there was a clear distinction between what "is" and what "should be." He argued further that scientific inquiry should be strictly confined to the former. For many of the contributors here, our PhD training also held out the supposed separation between positive and normative claims as legitimate and worthy of pursuit, in line with Friedman's preferred approach.

But, as our research has evolved and we have studied policymaking more closely or become involved directly, simplistic notions of keeping scientific inquiry separate from values have unraveled. Paradoxically, some of the same norms that support responsible engagement by elevating a particular kind of objectivity—including the exploration of alternative explanations and specification of study limits—are at times neglected by academics despite their training. Researchers' engagement in policy debates can amplify some of the most problematic impulses, including overgeneralization and the neglect of competing perspectives. In other instances, however, it is the striving for objectivity that is the core problem, as researchers conceal their political commitments, thereby misleading publics.

George DeMartino's chapter, "Should Economists Deceive? Prosocial Lying, Paternalism, and the 'Ben Bernanke Problem,'" examines the conditions under which economists feel justified in abandoning honesty and transparency with respect to what they can and cannot know. This is a clear departure from norms around scientific rigor associated with the economics profession. But DeMartino argues that the profession has normalized "prosocial lying"—that is, making false or unknowable statements about probable outcomes to shape public behavior. The wide acceptance among economists that prosocial lying or strategic speaking serves the public interest has made the practice a norm—albeit one that is antithetical to, yet coexists alongside, the academic enterprise. As DeMartino points out, however, many economists are so secure in their sense of superior knowledge vis-à-vis the public, they do not see the contradiction. They further believe that moving the public toward broader acceptance of trade liberalization or curtailing fear during a financial crisis serves the greater good, regardless of whether it is achieved through neglect of distributional consequences or even through explicit falsehoods. In his strong dissent, DeMartino points out that prosocial lying is coercive and paternalistic at its core, undermining the quality of policy debate and damaging confidence in expert opinion.

Rachel A. Epstein's chapter on "The Problem of Overgeneralization: The Effects of NATO Enlargement" also examines disciplinary norms and their effect on the quality of policy discourse. The debate over NATO enlargement reaffirms the value of exploring alternative explanations, study limits, and external validity to avoid overgeneralization—but also shows many violations of these desirable practices. Her chapter highlights the intrinsic difficulties of separating positive and normative claims, given that particular values are embedded within the theoretical and methodological choices taken by researchers, yet remain unacknowledged. This led researchers to claim they were arguing over the empirical results of NATO enlargement when in fact a different debate was also playing out. Under the radar, a core dispute was between those who valued democratic sovereignty of newly independent countries versus those who argued that concern for democratic sovereignty "could" never override the "realities" of so-called "great power politics." Epstein concludes, however, that scholars privileging the relationship between Russia and the US over the democratic sovereignty of East Central Europe might have replaced "could" with "should" in their defense of their "great power politics" approach, since elevating large countries' interests and perspectives reflects a political preference. Paradoxically, disciplinary norms discourage us from doing so. In striving for objectivity, scholars concealed their underlying political commitments, muddying the policy debate.

Understanding and Navigating Complex Policymaking Processes

This volume emphasizes scholars' options and duties to act in good faith, yet understanding and involving the policy side of the "gap" is also essential to responsibly inform policy. To engage effectively and ethically, forewarned is forearmed. Scholars need to understand the positions and motives of policymakers as well as when, where, and how they may be interceding in decision-making processes. With such knowledge, they can craft what information they share and how, so that it both makes more impactful contributions and avoids risks of being misused. Policymakers also have their parts to play in supporting successful engagement. By increasing their own awareness of the risks of politicization of findings, they can help ensure that research is interpreted and applied as intended. By learning about scholars' motivations and capabilities, they can also enter into interactions with researchers with better-calibrated expectations—about scholars' capabilities for analysis, and about their ethical boundaries and redlines.

In his chapter, "Data Ethics and the Power of Data," César Caballero Reinoso provides his perspective as both a practitioner and a generator of data who spanned analytical and policy roles both in and out of government. A former director of the Colombian census bureau (DANE) and a pollster, Caballero reflects on the importance of data and decision-making transparency and standing one's ground in the face of attempted seduction by—and pressure from—powerful political actors. Caballero describes how he set and adhered to clear moral standards and redlines which he would not cross, even while coping with risks as extreme as death threats, among other forms of retaliation. When he decided to quit his government position to avoid compromising his values, he started a survey company and continued to follow his ethical north star. Working with his employees, he veered into "partnered engagement" by developing an inclusionary and participatory process for them to collectively decide on their preferred research topics and policy engagement approaches. Caballero's experience also serves as a source of inspiration: if he could make responsible engagement choices under such real-world and personal pressures, then scholars facing less intense dynamics during their research and engagements can do so as well.

Leanne Erdberg Steadman, a former US government policymaker and national security staffer, contributes the final contributor chapter of the volume, "Policymakers' Epilogue: From the Receiving End of Responsible Engagement." Writing from the policymaker side of the "gap," she reflects on the scholars' engagement case studies and draws from them to offer her own recommendations. She sheds light on the various players and steps involved in policymaking processes—some of which scholars may not be aware of. She shows how it is crucial for policy-engaged scholars to gain an understanding of the policy process, including how policymakers are likely to misuse or misinterpret academic findings. Such an understanding can help scholars in both effectively sharing their findings and treading cautiously, given the possible diverging incentives and preferences that some policymakers may hold for how research should inform decision-making. Erdberg Steadman's analysis does not stop with academics, however. She also places some of the burden for responsible engagement on the shoulders of policymakers such as herself. She directs additional suggestions to policymakers for how they can play more proactive and responsible roles in their own efforts to bridge the academic–policy gap.

Taking Responsible Engagement Seriously

Survey research shows that many scholars in international affairs are actively engaged in policy processes. Few things could be more gratifying for scholars than being able to apply their research in constructive ways with measurable improvements in outcomes and human welfare. And yet our contributors' experiences reveal how difficult that realization turns out to be. As we have navigated everything from social pressure and competing political agendas to power rivalries and counterproductive disciplinary norms, we have reflected back on what we wished we had known and thought through *before* we engaged. While at one point in our careers we might have assumed that particular findings should compel a specific policy, we are now more aware of what can go wrong in the process of policy engagement, and of all of the mediating forces that can run between knowledge and policy implementation.

By presenting a series of engagement dilemmas in context and on a personal level, we hope to better prepare researchers, and policymaker consumers of research, for the challenges of policy engagement. Engagement can be very direct if a scholar is seconded to a policy position. Or it can be further removed, through publication of research that weighs in on a policy debate. No matter the mode of engagement, there are more and less responsible pathways. This is true whether one is researching on one's own or whether, as is increasingly common, as part of a larger research team. Either way, knowing and registering one's ethical redlines in advance of embarking on research is essential.

The contributors make a number of suggestions for improving the quality of academic engagement. These include formulating and publishing one's ethical redlines, thoroughly investigating the politics of the issues at stake as well as the positions of policy interlocutors, and getting the fullest view possible of the point at which one's research intervenes in the policymaking process. Thinking through the power dynamics intrinsic both to research and to policy engagement is also crucial for responsible conduct. Finally, informal disciplinary norms that paradoxically run counter to the research enterprise can be hard to spot because, as participants in our fields, we are susceptible to the normalization of some irresponsible conduct, including overgeneralization, prosocial lying, and the neglect of plausible alternative explanations. From this point of view, greater attention to responsible policy engagement not only potentially enhances our policy interventions, but can cultivate a sharper critical awareness of our own scholarly work and how to improve it.

References

Avey, Paul C., Michael C. Desch, Eric Parajon, Susan Peterson, Ryan Powers, and Michael J. Tierney (2022) "Does Social Science Inform Foreign Policy? Evidence from a Survey of US National Security, Trade, and Development Officials." *International Studies Quarterly* 66(1): 1–19.

Barma, Naazneen H., and James Goldgeier (2022) "How Not to Bridge the Gap in International Relations." *International Affairs* 98(5): 1763–1781.

Berry, Marie E., and Milli Lake (2021) "On Inconvenient Findings." *The Duck of Minerva* (blog). https://www.duckofminerva.com/2021/01/on-inconvenient-findings.html.

Center for Systemic Peace (2020) "Global Trends in Armed Conflict." https://www.systemicpeace.org/CTfigures/CTfig03.htm.

Diamond, Emily, Thomas Bernauer, and Frederick Mayer (2020) "Does Providing Scientific Information Affect Climate Change and GMO Policy Preferences of the Mass Public? Insights from Survey Experiments in Germany and the United States," *Environmental Politics* 29(7): 1199–1218. https://doi.org/10.1080/09644016.2020.1740547.

Epstein, Rachel A. (2008) *In Pursuit of Liberalism: International Institutions in Postcommunist Europe*. Baltimore, MD: Johns Hopkins University Press.

Fisher, Max (2022) "Is the World Really Falling Apart, or Does it Just Feel That Way?" *New York Times*, July 12. https://www.nytimes.com/2022/07/12/world/interpreter-world-falling-apart.html.

Friedman, Milton (1953) *Essays in Positive Economics*. Chicago: University of Chicago Press.

Gani, Jasmine K., and Jenna Marshall (2022) "The Impact of Colonialism on Policy and Knowledge Production in International Relations." *International Affairs* 98(1): 5–22.

George, Alexander L. (1993) *Bridging the Gap: Theory and Practice in Foreign Policy*. Washington, DC: United States Institute of Peace Press.

Gheciu, Alexandra (2005) *NATO in the "New Europe": The Politics of International Socialization After the Cold War*. Stanford, CA: Stanford University Press.

Hendrix, Cullen S., Julia Macdonald, Ryan Powers, Susan Peterson, and Michael J. Tierney (2023) "The Cult of the Relevant: International Relations Scholars and Policy Engagement Beyond the Ivory Tower." *Perspectives on Politics* 21(4): 1270–1282.

Iskander, George (2022) "The Manhattan Project Shows Scientists' Moral and Ethical Responsibilities." *Scientific American*, March 2, 2022. https://www.scientificamerican.com/article/the-manhattan-project-shows-scientists-moral-and-ethical-responsibilities/.

Jasanoff, Sheila (1990) "American Exceptionalism and the Political Acknowledgment of Risk." *Daedalus* 119(4): 61–81.

Kirshner, Jonathan (2022) *An Unwritten Future: Realism and Uncertainty in World Politics*. Princeton: Princeton University Press.

Kruse, Kevin Michael, and Julian E. Zelizer, eds. (2022) *Myth America: Historians Take on the Biggest Legends and Lies about Our Past*. New York: Basic Books.

Lewandowsky, Stephan, Laura Smillie, David Garcia, Ralph Hertwig, Jim Weatherall, Stephanie Egidy, Ronald E. Robertson, Cailin O'Connor, Anastasia Kozyreva, Philipp Lorenz-Spreen, Yannic Blaschke, and Mark Leiser (2020) *Technology and Democracy: Understanding the Influence of Online Technologies on Political Behaviour and Decision-Making.* Luxembourg: Publications Office of the European Union.

Mair D., L. Smillie, G. La Placa, F. Schwendinger, M. Raykovska, Z. Pasztor, and R. van Bavel (2019) *Understanding our Political Nature: How to Put Knowledge and Reason at the Heart of Political Decision-Making.* Luxembourg: Publications Office of the European Union.

Maliniak, Daniel, Susan Peterson, Ryan Powers, and Michael J. Tierney, eds. (2020) *Bridging the Theory-Practice Divide in International Relations.* Washington, DC: Georgetown University Press.

Mayer, Frederick (2014) *Narrative Politics: Stories and Collective Action.* Oxford: Oxford University Press.

Pechar, Emily, Thomas Bernauer, and Fredrick Mayer (2018) "Beyond Political Ideology: The Impact of Attitudes Towards Government and Corporations on Trust in Science." *Science Communication* 40(3): 291–415.

Pielke, Roger A. (2007) *The Honest Broker: Making Sense of Science in Policy and Politics.* Cambridge/New York: Cambridge University Press.

Rathbun, Brian C., Caleb Sebastian Parker, and Caleb E. Pomeroy (2024) "Separate but Unequal: Ethnocentrism and Racialization Explain the 'Democratic' Peace in Public Opinion." *American Political Science Review.* Published online: 1–16. doi:10.1017/S0003055424000509.

Roser, Max (2018) "Twice as Long—Life Expectancy Around the World." *Our World in Data,* Oct. 8. https://ourworldindata.org/life-expectancy-globally.

Scharfbillig, M., L. Smillie, D. Mair, M. Sienkiewicz, J. Keimer, R. Pinho Dos Santos, H. Vinagreiro Alves, E. Vecchione, and Scheunemann L. (2021) *Values and Identities: A Policymaker's Guide.* Luxembourg: Publications Office of the European Union.

Schweber, S. S. (2000) *In the Shadow of the Bomb: Oppenheimer, Bethe, and the Moral Responsibility of the Scientist.* Princeton: Princeton University Press.

Stone, Deborah (2018) "The 2017 James Madison Award Lecture: The Ethics of Counting." *PS: Political Science & Politics* 51(1): 7–16. https://doi.org/10.1017/S1049096517001767.

World Bank (2016) *Poverty and Shared Prosperity 2016: Taking on Inequality.* Washington, DC: World Bank.

World Bank (2020) *Poverty and Shared Prosperity 2020: Reversals of Fortune.* Washington, DC: World Bank.

Zvobgo, Kelebogile, and Meredith Loken (2020) "Why Race Matters in International Relations." *Foreign Policy* no. 237 (Summer): 11–13.

2

Pre-register Your Ethical Redlines

Vulnerable Populations, Policy Engagement, and the Perils of E-Hacking

Jason Lyall

In a burst of introspection, scholars have recently sparked a welcome discussion about best practices for ethical fieldwork (Wood 2006; Fujii 2012; Campbell 2017; Cronin-Furman and Lake 2018; APSA 2020).[1,2] The stakes are high. Scholars now routinely conduct fieldwork among vulnerable populations, often in fragile or violent settings, and do so with sometimes large-scale interventions that can impose significant harm if they go awry.[3] In addition, a new generation of scholars is now conducting this fieldwork with an eye toward influencing policymakers, whether through direct collaboration or indirect channels.[4] Fieldwork and policy engagement have, for many of us, become intertwined (Oliver and Cairney 2019).

Yet several ethical dilemmas arise at the intersection of research and policy. Scholars at this nexus are perhaps uniquely vulnerable to crosscutting

[1] For helpful comments, I thank Deborah Avant, Erica Chenoweth, Dara Cohen, George DeMartino, Rachel Epstein, Cullen Hendrix, Oliver Kaplan, and Zoe Marks. I also benefited greatly from discussions with participants at the University of Denver's Responsible Public Engagement Institute (2021 and 2022) and the Advanced Workshop on Ethical Engagement in Conflict Research at the Weatherhead Center for International Affairs at Harvard University (9–10 July 2019), where it was originally presented. All errors are mine.

[2] The Advancing Conflict Research (ARC) bibliography provides an excellent overview of these materials (https://advancingconflictresearch.com/arcbibliography)

[3] I define a vulnerable population broadly as individuals or groups that live in violent, fragile, or nondemocratic settings where (1) participants have experienced harm or credible threat of harm; (2) participants are at risk of further harm, whether mental, physical, or economic; and (3) proposed fieldwork might increase risk of harm (e.g., via retaliation). Examples of populations include traditional categories such as children, pregnant women, and inmates (National Committee for the Protection of Human Subjects of Biomedical and Behavioral Research 1979), but also extends to refugees, internally displaced persons (IDPs), and migrants. In some cases, the extent or threat of violence by governments, armies, or insurgents might be so high that the entire population of a fieldwork site would be defined as vulnerable. On ethical dilemmas in large-scale field experiments, see McDermott and Hatemi 2020.

[4] See, for example, the important Bridging the Gap initiative (https://bridgingthegapproject.org/) and the Carnegie-funded Responsible Engagement of Policymaker initiative at the University of Denver (www.sieethicalengagement.com/).

Jason Lyall, *Pre-register Your Ethical Redlines.* In: *Speaking Science to Power.* Edited by: Rachel A. Epstein and Oliver Kaplan, Oxford University Press. © Jason Lyall (2024). DOI: 10.1093/oso/9780198875185.003.0002

pressures from local populations and policymakers that can lead to unexpected and unwelcome ethical compromises—what I call "ethics hacking" or "e-hacking"[5]—to produce policy-relevant scholarship. These ethical issues are largely structural in nature: regardless of our preferred methodological stance or fieldwork setting, all scholars face a similar set of dilemmas arising from the challenges of lashing fieldwork to the mast of policy relevance.[6]

We therefore need a structural approach for identifying and addressing possible ethical dilemmas before conducting fieldwork and policy engagement. While few scholars set out to violate ethical norms, the collision of competing research incentives, tight deadlines, a desire to shape policy, and fast-moving dynamics on the ground can lead to piecemeal, and unintended, ethical compromises. Taking a page from the preregistration movement (Humphreys, Sanchez de la Sierra, and van der Windt 2013; Nosek et al. 2018), I argue that scholars should pre-register their ethical redlines. That is, we should each publicly designate the types of action (or inaction) that we deem to be ethically out of bounds when engaged in fieldwork among vulnerable individuals and communities. More specifically, a preregistration document captures our bedrock ethical principles, how we assess risk and potential harm to research participants, and how we propose to share our findings with policymakers and the broader public.

Preregistration has many benefits. It provides a mechanism for the transparent accounting of our risk assessment as well as steps for harm mitigation that go well beyond standard Institutional Review Board (IRB) practices. When made publicly available, preregistration can help defuse e-hacking pressures by serving as a credible commitment to one's research ethics.[7] In fact, preregistration acts as a signal of intent that can screen out research partners with questionable ethical practices, helping avoid ethical dilemmas from arising in the first place. Written before fieldwork, a preregistration document forces scholars to consider the end-use of research findings, how results will be conveyed to decision-makers, and possible misuse by other actors. Preregistration is compatible with all methodologies, including participant observation, interviews and focus groups, surveys, and field experiments. And, with sufficient buy-in among scholars, we might create norms that are

[5] I define e-hacking as the piecemeal shifting—sometimes unnoticed, sometimes calculated—of a scholar's prior ethical boundaries on the conduct of fieldwork and subsequent use of research findings.

[6] Desk research can also raise ethical concerns, particularly surrounding exposure to "second-hand trauma." See, for example, Hoover Green and Cohen 2021. I focus here on the intersection of fieldwork, vulnerable populations, and policy engagement since the ethical trade-offs and possible harms are most stark in these settings.

[7] On credible commitments, see Fearon 1997.

powerful enough to shape expectations of appropriate conduct in the absence of a discipline-wide ethical code.

This chapter sketches how we might pre-register our ethical redlines for fieldwork and policy engagement. I begin with a discussion of common sources of ethical fieldwork dilemmas, especially during collaborative projects, and detail the forms that e-hacking can take. I then outline the benefits of public preregistration before turning to a discussion of how we can formalize the writing of these documents. In effect, preregistration forces us to consider the risks and potential harm of our research across five audiences: researchers, team members, individual participants, host and adjacent communities, and policymakers, including the broader public. I conclude with eight concrete steps that we can take to protect our ethical redlines. I draw on my experiences conducting fieldwork in Russia and Afghanistan, where I have worked with government agencies (the US Air Force, USAID, the US Army), nongovernmental organizations (Mercy Corps, IRI), survey firms (Orca, ACSOR), human rights activists (Memorial), and ethics boards (US Special Forces Command, University of Denver's Responsible Policy Engagement initiative), among others.

What's at Stake: Why We Need Responsible Policy Engagement

Seeking policy relevance, scholars and their collaborators have unleashed a surge of ambitious research with clear, if sometimes unacknowledged, ethical implications for vulnerable populations in fragile settings. A small sample of recent research highlights this trend. Researchers, for example, have randomly incentivized Hong Kong university students to attend anti-authoritarian protests against China to explore why individuals participate in high-risk activism (Bursztyn et al. 2021). Randomized exposure to videos of actual violence have been used to trigger group identities in war-torn Kashmir (Nair and Sambinis 2019). Citizens in Cali, Colombia, were randomly exposed to militarized police patrols that led to increased reports of crime and human rights abuses in the weeks following the intervention (Blair and Weintraub 2021). Efforts to raise awareness of corruption in Lagos, Nigeria, appear to have backfired, with some individuals reporting a greater willingness to pay bribes (Cheeseman and Peiffer 2022). High-profile ethnographic work in Afghanistan has been challenged for failing to safeguard the anonymity of local sources, leading to retaliation (AAA Commission 2009). Large-scale efforts by investigators to encourage the use of ride-hailing services like Uber and Lyft have boosted, not decreased, traffic congestion and pollution in American cities, especially within poor communities (Diao, Kong, and Zhao 2021).

There is no question that these researchers are asking important, policy-relevant questions. But the stakes are high; the potential for widespread harm is also embedded within many of our research questions and designs. Field teams and survey respondents alike might face retaliation or social harm (such as being ostracized), or suffer reawakened trauma by participating in research studies. Local conditions, whether measured in terms of crime, economic opportunities, or intergroup trust, might be worsened in the days, weeks, and months following these efforts. Broad societal dynamics, including trust in authorities and decisions over public spending, might also be distorted by our research interventions. At worst, our efforts might create or accelerate intergroup competition and societal conflict, sparking new rounds of violence. Standard IRB protocols are unlikely to prevent, or even notice, these potential societal harms. And that such harms are unintended or unexpected does not absolve researchers of the need to consider what ethical obligations they have to local collaborators and populations, or how to minimize any risks and negative externalities to broader communities.

These stakes are raised even higher by increasingly close collaboration between scholars and policymakers.[8] A bridge, albeit sometimes shaky, now connects academics to decision-makers in both governmental and nongovernmental agencies seeking robust evidence for policy design and implementation. In many cases, the growth in our empirical ambitions has been matched by the rise of a research infrastructure that facilitates conducting truly large-scale, sometimes even global, interventions. If our research is not ethically grounded, we risk inflicting widespread harm on participants and their communities. This is true even if researchers work alone or in isolated settings if findings from unethical research are injected into the policy process. In short, even if our research endeavors are perfectly executed, the scale and ambition of current efforts to influence policy should lead us to consider the strength of existing ethical guardrails.

What we need, then, is responsible policy engagement. Regardless of methodological approach, I view responsible engagement as having several qualities. Above all, the fieldwork that informs policy recommendations must

[8] This collaboration may be formal, where scholars embed themselves in partner organizations; more distant, where scholars join with fellow scholars, teams, and organizations to undertake research to analyze specific policy problems; or arm's length, where fieldwork is conducted independently but with an eye toward informing debates about specific policies. Some scholars will move between these forms of collaboration on a project-by-project basis over the arc of their careers.

minimize the prospect of current risk and future harm to participants, their communities, and adjacent populations. The empirical strategy used to generate these insights should be rigorous, meaning that key decisions about sampling, data collection, and analysis are transparent and appropriate for a given empirical setting. Both the instruments (interviews, focus groups, surveys, randomized interventions) and nature of analysis should be preregistered. Samples should be large enough to support claims; care should be taken not to extrapolate results to dissimilar contexts (see Epstein, Chapter 7, this volume). Findings and their limitations, along with any weaknesses in design or analyses, should be discussed openly rather than buried in an appendix (or ignored completely). To guard against mistaken inferences, findings should be peer reviewed or audited by independent actors before being injected into the policy process. Alternative interpretations should be engaged fairly when advocating for a policy position, especially if that recommendation is at odds with prior scholarship. Mistakes will still occur, of course. But we should at least commit to making a good faith case for our preferred policy, one grounded in both ethics and evidence.

The Drivers and Perils of E-Hacking

The danger of e-hacking—of making ethical compromises—looms when we aim to wrench policy-relevant insights from fieldwork in fast-moving and uncertain settings under tight deadlines. More specifically, there are at least three drivers of e-hacking.

Most simply, scholars may misjudge the risk and potential harm of proposed feedback and take ill-advised actions to save their research agenda. Many ethical dilemmas arise from a poor understanding of the baseline risks facing a given population. To date, we have yet to normalize the reporting of pre-study risk assessments in our published work. As a consequence, we lack established frameworks for judging the thoroughness (and realism) of our pre-fieldwork assessments. Since we lack a shared set of assessments, we cannot learn from others' experiences. We therefore lack established pre-study benchmarks for determining when our fieldwork might be sliding off the rails. In addition, local dynamics can create new, unanticipated, ethical dilemmas. Our own research designs might inadvertently lock us into situations where we are generating risks to participants. Researcher inexperience can compound these difficulties. While all researchers are vulnerable to mistaken risk assessments, scholars entering a new setting are likely to miss important local trends before (and even after) they arrive. Unprepared, these scholars might find themselves entrapped into unanticipated situations that force ethical compromises that they would ordinarily denounce. Recasting, or even abandoning, fieldwork can be daunting; powerful incentives exist to stay the course, or even double down, in the face of these initial mistakes.

Genuine affection for an organization or its mission can also create pressures to shift ethical redlines. The chance to play an outsized role in the policy process in an important struggle—say, for example, encouraging protests against an autocratic regime, or supporting humanitarian aid in war-torn countries—can be exhilarating. But it can also produce distorting biases that cloud judgment.[9] Ethical safeguards can quickly be viewed as cumbersome obstacles that must be surmounted if timely action or policy recommendations are required. Pro-organization beliefs might also lower inhibitions on the use of dubious data. Peer review, quality control, and sensitivity checks might all be sacrificed in the service of the mission (the "greater good"). Close affinity with an organization's mission can also warp research questions themselves; certain data analyses might be abandoned in favor of others deemed more relevant by policymakers. Taken too far, these biases can overturn one's independence as a scholar, calling into question one's findings and undermining the fieldwork's original purpose. Having sacrificed intellectual freedom, we might find ourselves bereft of our scholarly autonomy while saddled with biased, and perhaps simply empirically wrong, policy advice.

Sometimes ethical lapses are driven by naked careerism. Material rewards—whether money or early access to proprietary data—are an obvious possible inducement to cutting ethical corners.[10] Researchers might feel pressure to down-weight inconsistent or inconvenient results to produce a cleaner narrative. Nuance might be jettisoned. Weaknesses in the research design, or implementation problems, might be hidden in appendices, if not buried completely. Initial reluctance to conduct "off-book" investigations outside normal peer review channels might give way if these become the price for continued data access. Such "off-book" analyses are especially problematic if they are done under time pressure and with an eye toward score-settling

[9] Anthropologists use the somewhat unfortunate term "seduction" to capture the unconscious process by which a researcher comes to identify so closely with participants or organizations that objectivity is lost. See, for example, Devereaux 1967; Nordstrom and Robben 1996.

[10] As Zingales 2014 notes, proprietary data can be instrumental for publishing in top-flight journals, especially for early career academics, creating incentives for organizations and companies to dangle data access to "capture" academics. For a particularly graphic example of how cash and data can create bias and e-hacking pressures, see "Uber paid academics six-figure sums for research to feed to the media," *The Guardian*, 12 July 2022, www.theguardian.com/news/2022/jul/12/uber-paid-academics-six-figure-sums-for-research-to-feed-to-the-media.

within partner organizations. As such, they can become quickly politicized and injected directly into the policy process without the normal safeguards that might induce caution.

E-hacking might also encourage the appearance of its close cousin, p-hacking: the use of repeated estimation strategies to "discover" statistically significant results that support desired research outcomes (Humphreys, Sanchez de la Sierra, and van der Windt 2013). Researchers might also find themselves implicated in the use of unreliable data, sometimes gathered through ethically dubious means, to maintain a collaboration. In some cases, human subjects' protections might also be relaxed to facilitate a partner's "mission critical" work among vulnerable populations.[11]

Faced with a slow drip-feed of concessions to partners, researchers can find themselves reluctantly going along to protect their future income and publication streams. Collaborations can also turn coercive. Power imbalances are often a hallmark of scholar–practitioner partnerships. Reliance on these partners for shelter, movement, and security in difficult field settings reinforces these power asymmetries. As these dependencies grow, so too does the potential for unwelcome pressure on ethical standards. Financial dependencies can be especially tricky to navigate. If scholars are not bringing resources to the table, either in the form of money or desirable technical skills, then they may find themselves overruled in disagreements over the research agenda or its implementation. Without skin in the game, so to speak, your ethical concerns might be dismissed. Sensitive topics or particular field sites might be tabled for fear of disrupting the relationship, especially at its early stages (see Olsen, Chapter 3, this volume). Scholars might also be compelled to shift their hypotheses and research designs to accommodate organizational sensitivities. These compromises not only restrict intellectual freedom but can also weaken safeguards meant to reduce risk and harm among vulnerable populations. The need to demonstrate favorable initial findings is especially likely to generate these coercive and corrosive pressures.

As this discussion implies, e-hacking is not typically a one-time event. Ethical concessions can spark new fieldwork challenges that, in turn, require additional compromises. Prior estimates of risk and potential harms might be continually revised downward to justify the continuation of problematic fieldwork. Scholars might become overly deferential to their partners, willing to mold their research around organizational sensitivities and priorities rather than theoretical or empirical debates. Critical voices on a research

[11] Organizational environments where turnover is high among staff might also encourage ethical shortcuts. The need to demonstrate one's worth to new, possibly skeptical, decision-makers quickly can create misaligned incentives between careful and rapid findings.

team can be discounted, or even silenced, if a member is seen as jeopardizing the relationship with a partner organization ("not a team player"). Homogenous teams are also more susceptible to e-hacking since they lack the diversity that serves as an internal check on ethical lapses.[12] Without a robust mechanism for arresting this slide, researchers risk creating a cascade of ethical concessions from initial, perhaps small, missteps, unraveling the integrity of their fieldwork and its policy conclusions.

I am not suggesting that all potential partners are unscrupulous or seeking to take advantage of scholars. Far from it. Many productive collaborations have been forged between academics and organizations such as Evidence in Governance and Politics (EGAP), Mercy Corps, the Abdul Latif Jameel Poverty Action Lab (J-PAL), the International Rescue Committee (IRC), amongst others. And it is possible that scholars might use their own expertise (or reputation) to browbeat research partners into unethical practices, particularly if future donor funds are tied to program effectiveness. Most of the time, however, it is the researcher, not the organization, with the weaker hand. Even if no coercion is intended, there are enough power asymmetries, fast-changing conditions, and donor demands associated with fieldwork to create incentives to e-hack. Where concessions are often piecemeal in nature, staggered over multiple decision points, and individually (somewhat) reasonable, scholars may belatedly realize that they have compromised their ethical principles while chaining themselves to research projects built around the external partner's needs and ethical standards.

Combating E-Hacking through Preregistration

Preregistration offers one means of arresting the slide into situational ethics. I define preregistration as the act of publicly documenting a researcher's ethical code, their best assessment of the potential risk and net benefit of the proposed research to specific communities, and harm mitigation efforts that will guide the investigation and dissemination of findings before research is actually conducted. This document should be formal and extensive. That is, researchers should take the opportunity, in the quiet before the storm of fieldwork and policy engagement, to consider their ethical redlines.[13]

To maximize the advantages of preregistration, these documents should be public, for three reasons. First, a public statement acts as a credible

[12] On the advantages of diverse teams, see Page 2017.
[13] Some scholars have already been writing their own personal codes of conduct (see, for example, Chenoweth 2020). Just how common this practice is remains unknown.

commitment device. Identifying your redlines and publicly pledging to conduct ethical fieldwork allows other researchers to assess compliance with stated ethical principles. The prospect of being evaluated can generate external pressure to avoid e-hacking even in situations where the researcher faces no oversight—a common feature of fieldwork in violent or fragile settings. This external pressure can be helpful, acting as another line of defense against demands by research partners to engage in conduct that would compromise one's values. For example, the document's public nature, along with the need to provide detailed justification for deviations (or breaches) of these redlines, creates a shield that can deflect attempts to impose new requirements or force last-minute changes that might produce unethical research. In particular, unwelcome pressure to conduct side analyses could be short-circuited by a public preregistration document.

Second, preregistration signals a researcher's expectations to prospective partners about how the fieldwork will be conducted. By shaping expectations early, scholars might avoid inadvertently selecting themselves into collaborations where ethical compromises become necessary. Unscrupulous partners, or those with clashing ideas of ethical conduct, may simply steer clear of scholars who make their redlines explicit. If enough scholars commit to public preregistration, bad actors might find themselves deprived of research partners unless they adjust and improve their own fieldwork practices.

Third, public preregistration represents something of a collective good. By making ethical preregistration statements available, interested researchers, especially those without substantial experience in difficult settings, can craft their own ethical standards in dialogue with others. New questions and possible answers are likely to arise when reviewing others' ethical standards. Such a review helps scholars anticipate possible problems with fieldwork and policy engagement, creating space for a careful consideration of possible solutions. It also provides a springboard for discussions of how to conduct a thorough risk assessment and harm mitigation by facilitating comparison with earlier efforts (Baron and Young 2021). In short, if these documents are made public, reinventing the wheel is not necessary each time a researcher enters into collaborative fieldwork.

Statements of ethical conduct could easily be folded into existing preregistration templates provided by EGAP or the American Economic Association (AEA).[14] The open-source software DeclareDesign could also be easily

[14] EGAP now requires preregistration with the Open Science Foundation (https://osf.io/prereg/). The AEA registry can be found at www.socialscienceregistry.org/. Neither currently requires ethical statements as part of preregistration.

updated to include a section on risk assessment and harm mitigation.[15] Scholars might also choose to append their statements to IRB submissions or post them as standalone documents on their own website. Whatever the format, it is preferable to have the ethics declaration and research design overview in the same document. This reinforces the dialogue between design and ethics, underscoring that many design issues are deeply imbued with ethical implications.

Locating Redlines: General Principles

Ethical redlines mostly stem from two sources: researchers' own bedrock ethical principles about what constitutes appropriate fieldwork and policy engagement, and context-specific challenges that arise from fieldwork in fast-moving settings.[16] No one can fully anticipate all possible challenges: personality clashes, competing organizational imperatives, new information, and sometimes sudden changes in local environments all frustrate a one-size-fits-all approach. Yet while trying to anticipate everything is a recipe for paralysis, our inability to anticipate every contingency is not a reason to avoid preregistration.

One therefore needs to locate their ethical redlines. A preregistration exercise offers one vehicle for identifying core ethical principles and for articulating the possible risks and harms inherent in the proposed research. Painting with a broad brush, we must engage in an honest assessment of the nature and magnitude of risks to multiple audiences; outline possible short- and long-term harm among participants and communities, including the poststudy dissemination of findings; and identify possible measures that might ameliorate these harms. Well before research designs are crafted, researchers should explicitly theorize the proposed intervention's direct and indirect effects on both the intended and any adjacent populations.

Central to this task will be articulating who these populations are, possible spillover channels, and the potential benefits and harm that might result for (non)participating communities. It is necessary, in other words, to render explicit the implicit causal model that guides the proposed research. How might the proposed research affect these populations? (see also Epstein,

[15] The open-source software DeclareDesign enables preregistration of the scholar's model of the world (M), an inquiry (I), a data strategy (D), and an answer strategy (A). It could be extended to include ethical redlines (E). See Blair et al. 2019.

[16] Institutional partners and collaborators rarely have their own formal codes of ethical conduct. These organizations do often have informal norms surrounding appropriate research behavior that can influence the location of ethical redlines.

Chapter 7, this volume) What is the worst-case scenario for harm among these vulnerable populations? What are the relevant counterfactuals for the proposed research, and do alternative questions or methods exist that might reduce anticipated risks? Some interventions will be intrusive;[17] others, more subtle;[18] and still others confined only to the researcher's own status as an outsider in the local context.[19] All fieldwork, however, has the potential for ethical missteps, creating a demand for preregistration even when goals and methods appear modest.

The following subsections offer some general scaffolding for building a pre-registration document. I focus on five specific issues: (1) the researcher's own bedrock principles; (2) the role of collaborators, including local partners; (3) the treatment of research participants; (4) consideration of group rights, including both host communities where research will be conducted and adjacent ones that might be indirectly affected; and (5) responsible engagement with policymakers and the broader public.

Your Bedrock Principles

Ethical redlines are reflections of deeply personal bedrock principles. No two scholars may agree on where to draw their boundaries, but each of us is guided by a sense of ethical behavior that shapes our fieldwork and policy engagement. This intuitive sense of appropriate conduct, of right and wrong, provides the foundation for our preregistration documents.

To begin, field researchers must take inventory of their own risk profiles before engaging in fieldwork. What types of physical and, equally as important, mental risks and challenges are researchers willing to assume by engaging in the proposed fieldwork? Everyone has their own tolerance for risk, and this can change over time. Individual senses of personal risk have downstream consequences for how one designs and conducts fieldwork, so we must be explicit about what we consider to be necessary and unnecessary risks. How much risk is too much, for example, and how comfortable is one asking others—collaborators, colleagues, research participants, host communities—to assume those risks? Are risks to personal health

[17] I have in mind interventions that seek to change existing political, economic, or societal arrangements. These might take the form of large-scale vocational training, cash transfers, or police reform programs.

[18] These might include survey experiments, encouragement designs to facilitate collective action (i.e., to encourage voter turnout), or informational primes about intergroup relations.

[19] See Cilliers, Dube, and Siddiqi 2015 on how outside status can affect research results. It may also raise ethical concerns if outsider status raises expectations of (continued) material assistance for participating in the research.

shared equally, or are the researchers asking others to run risks on their behalf?

We must also consider the possibility that the researcher's own presence can increase the risk of harm to others. Thinking through the nature of those negative consequences and their possible direction—does a researcher's presence upset existing power or social hierarchies, for example—is a key component of risk assessment. In some cases, the researcher's initial intervention might have far-reaching but slow-developing consequences, forcing us to consider not just the immediate effects of our research but also the downstream risk. Part of this risk calculation should be the weighing of alternatives: is fieldwork really necessary, or are there alternative means of collecting evidence that lower the risk to others? In some cases, the consequences of an inadvertent breach of data confidentiality may be so severe for participants that ethical fieldwork is impossible. Here, the presence of a researcher and the sensitivity of collected data collide with emerging norms about transparency and replication in the social sciences.[20] Where the researcher stands on the trade-offs between transparency and risk management will be important to pre-specify. Scholars should also consider whether certain populations are simply too vulnerable to conduct ethical fieldwork within.

Collaborative research and policy engagement also raise the risk of a loss of intellectual freedom. Scholars should use their preregistration exercise as an opportunity to consider how research partners might chip away, sometimes unknowingly, at the ability to ask and answer your research questions (see Olsen, Chapter 3, this volume). As noted earlier, resource imbalances, tight timelines, and a desire to maintain a relationship can all lead scholars to make concessions that might compromise intellectual freedom. Some scholars are comfortable working on a research project devised by an external organization; others will want to craft their own research questions and then find implementing partners. Some collaborations blend these two ideal types. In each case, it makes sense to locate the limits of one's willingness to cede control over (parts of) the research process to outside actors and to craft possible safeguards for preserving intellectual freedom.

These collaborations can also blur the responsibility for oversight of the research and its potential consequences. For example, I conducted a pre-registered field experiment in Kandahar, Afghanistan, that involved a Western NGO, six Afghan ministries and agencies, two international organizations, a local telecommunications company, a survey firm, two colleagues at different universities, and three different donors (Lyall, Zhou, and Imai

[20] See, for example, the discussions among political scientists about the Data Access and Research Transparency (DART) Initiative (APSA 2020).

2020). In these complex settings, it is crucial to identify the researcher's specific responsibilities, including who is accountable if harm is inflicted, and who holds final authority for the design and conduct of fieldwork. In these settings, preregistration prompts researchers to consider their general comfort with the delegation of responsibility. It also offers a blueprint for ensuring that all parties have clear expectations about their responsibilities (and accountability) when working among vulnerable populations (Figure 2.1).

Finally, we must define the terms of responsible engagement with policymakers. Part of this process is simply identifying the organizations that one will, and will not, work with. This can be difficult in practice: researchers must often collaborate with repressive or corrupt actors, including governments, nongovernmental organizations, and companies, as the price of access to vulnerable populations. Similar questions surround research funding: are some donors ethically off-limits? Working with certain organizations, especially militaries and private companies, raises ethical questions around whether scholars are willing to use classified or proprietary data that cannot

Figure 2.1 Survey team conducting fieldwork measuring public attitudes in a village in Helmand, Afghanistan, 2010.

Photo: Jason Lyall

be shared with other researchers. Whether these data were acquired or collected ethically raises "second-hand" dilemmas even if a researcher is not involved in the actual data collection.[21]

Much of this discussion will hinge on the researcher's desired role in the policy process. Some will prefer an embedded role; others, a more distant collaboration; and still others will be most comfortable with an arms-length, more indirect role in shaping policy debates. The same holds true for the scholar's public profile. What does "success" mean in this space, for example, and what's the best means of obtaining it? Some scholars will be comfortable with public-facing scholarship that paints in broad strokes and smooths over (or ignores) the nuances in the findings. Others, however, might relish these nuances and caveats, seeking to couch findings in degrees of uncertainty even if it means reducing one's audience (see Sisk, Chapter 4, this volume). All researchers conducting public-facing research will need to wrestle with the question of their desired role in policy debates, including how to conduct ethical outreach on social media and other popular outlets long after the fieldwork is completed.

Research Partners

Standard IRB protocols exclude consideration of the safety of our research partners and, in particular, the individuals and teams that make fieldwork possible. This omission is unfortunate: reliance on "research brokers" (Baaz and Utas 2019) raises a host of questions that should be explicitly folded into our ethical frameworks. Perhaps most importantly, researchers should assess the risks that local fixers, translators, research assistants, enumerators, confederates, and other teams face as a result of their involvement in fieldwork.[22]

What measures, for example, do you need to take to mitigate these risks? What types of harm—economic, reputational, physical—are most likely in these settings? In a post-pandemic world, can health risks be mitigated? And what of the mental strain of the fieldwork itself? Issues surrounding equitable pay (and what constitutes a fair wage) and credit-sharing, including coauthorship, should also be addressed in the preregistration document. Similarly, researchers need to think through what obligations they hold toward their team members if they are hurt (or worse) while conducting fieldwork.

[21] Here I have in mind interviews and surveys among captured insurgents (or suspected insurgents) who likely suffered physical and mental abuse.

[22] Desk research can also raise ethical concerns, particularly surrounding exposure to "second-hand trauma." See, for example, Hoover Green and Cohen 2021.

Collaborative fieldwork also raises the issue of differing ethical standards among team members. How to reconcile competing visions will remain an ongoing challenge for teams if these differences remain unacknowledged and unaddressed before fieldwork begins. Your ethical redlines, along with proposed solutions for mitigating ethical dilemmas, should be part of any group preregistration exercise. Many of the pressures that lead to ethical shortcuts come from within-group dynamics rather than external partners or the dynamics of conflict settings. This is true, too, of donors who view themselves as part of the research team and that have differing ethical standards. In a worst-case scenario, varying ethical frameworks between a lead researcher, team members, and external organizations might generate substantial pressure to engage in e-hacking just to maintain the collaboration. While the ethical codes of others are often blind items, we should be diligent about gauging their redlines (if any), including their understanding of the risks and harms associated with the proposed research.

Finally, the issue of deference often arises in fieldwork. Should local partners carry a decisive veto over risk assessments and how to respond to adverse events? How much weight should be given to local preferences over site location or interview subjects? What happens when there is disagreement within the team over design issues or policy recommendations? And how much autonomy will team members have for adjusting to adverse events, including changing field sites or interviewees, once in the field? While we tend to defer to local expertise, team members have their own identities and interests. These can sometimes work at cross-purposes to the proposed research and create new ethical dilemmas. In Afghanistan, for example, local views often carried enormous weight in planning and delivering aid projects. In some cases, however, this led to aid diversion, the rise of corruption, and sabotaged impact evaluations, leaving researchers unable to gauge aid's effects on local attitudes and violence. Team dynamics, and how to manage possible ethical dilemmas arising from them, should be addressed in our preregistration documents.

Individual Participants

Much has already been written about the ethical treatment of research participants drawn from vulnerable populations in (post)conflict settings.[23] Here I would simply reiterate that we must move beyond the legalistic protections afforded by IRB protocols to consider broader notions of ethical research. In

[23] See again the ARC bibliography: https://advancingconflictresearch.com/arcbibliography.

particular, the issue of informed consent looms large. Preregistration exercises are an ideal venue for defining what constitutes "informed consent" among these populations and how it will be obtained. How can one ensure, for example, that participants not only understand the initial decision to offer consent, but that they retain the right to revoke it at any time? How much deception, if any, is ethical under these conditions? What happens if interviews and surveys, to take two examples, are public in nature? Is informed consent even feasible if a site is controlled by a repressive government, criminal organization, or rebel group? What does informed consent mean when access to a location is brokered with a rebel (or criminal) organization or repressive state agencies?

Scholars also need to be proactive about considering the risks of retaliation against participants. Given ever-increasing powers of state (and nonstate) surveillance, researchers need to document the nature of the precautions they will take to preserve participant anonymity and the confidentiality of their responses or behavioral data. Studies that encourage high-risk activism—say, participation in mass protest in autocracies—or that solicit views on sensitive topics, such as support for criminal or insurgent organizations (or, indeed, the government), need to be especially careful to build robust data protection measures. Direct questions, for example, might be abandoned in favor of indirect questioning techniques that obscure the respondent's identity. In some cases, the risk of retaliation might prove too high to conduct ethical research. The Taliban capture of enormous quantities of personally identifiable information on aid recipients from humanitarian aid organizations after the fall of Kabul in 2021 is a stark reminder of the dangers of lax data security (Loy 2021).

Preregistration can also be helpful in detailing how participants might be harmed by their involvement. Harm can take many forms. Retraumatization, for example, is a possible danger for individuals who have been victimized in these settings. Given this possibility, certain aspects of an individual's personal history might be off limits or require an indirect approach. Similarly, partnership with aid agencies or nongovernmental organizations might stoke hope among individuals of future material rewards for participation. These expectations might have distorting effects on the research; they may also prompt backlash from angry individuals who believe that they have been misled or exploited. There may also be invisible harms, including loss of wages, disrupted studies, reputational loss, or other opportunity costs borne by participants.

One additional issue deserves mention. Every study makes an implicit trade-off between the potential costs and benefits of the proposed research. There is a danger, however, in fieldwork that draws on too few individuals to make a meaningful contribution about local populations, let alone more generalizable claims. We need to think hard about how to right-size our samples to avoid both needless harm by scaling up too much but also the risk that our research is too underpowered to make credible claims. Studies that are too small to make credible claims but that still heighten risks among vulnerable populations represent the worst of both worlds. We must be mindful of the possible risks facing participants while avoiding the trap of too-small samples that might lead to mistaken inferences.[24]

Host and Adjacent Communities

Preregistration also creates the opportunity to consider one's ethical obligations toward host and neighboring ("adjacent") communities. We often treat spillover between individuals (or villages) as a methodological problem to be mitigated, if not eliminated completely. But our research can also create ethical spillover if its consequences are felt within host communities or ripple beyond them.

Our calculations of risk and potential harm should be extended to the communities that host our research, rather than focusing narrowly on participants only. Informed consent, for example, may involve seeking formal permission from local power-brokers to conduct the desired research. Our interventions, and, indeed, our mere presence, can often upset existing power or social hierarchies in host communities. Economic interventions, for example, may create a new set of winners and losers within host communities that can have long-term effects, including decreasing trust, stoking jealousies, and creating incentives to engage in violence.[25] In short, what are scholars' short- and long-term obligations to the host communities that make the fieldwork possible?

[24] For example, there is a growing trend in conflict studies of using small-n samples to survey attitudes among vulnerable populations. While these might be motivated by a desire to minimize potential risks, they may be too underpowered to generate credible causal inferences. Risk is therefore imposed on these populations without producing useful knowledge.

[25] This is true even of household-level interventions, where family and gender hierarchies may be scrambled by participation in vocational training or cash transfer programs.

One of the most important—and, to date, neglected—areas of ethical concern is that of group rights and demographically identifiable information (DII). Following Raymond (2017), DII is

> either individual and/or aggregated data points that allow inferences to be drawn that enable the classification, identification, and/or tracking of both named and/or unnamed individuals, groups of individuals, and/or multiple groups of individuals according to ethnicity, economic class, religion, gender, age, health condition, location, occupation, and/or other demographically defining factors.

Nearly all ethical discussions focus on the researcher's own data collection (a "single stream" approach). We must be aware, however, of how our data can be merged with other, usually independent, fieldwork to create a new picture of host communities that might increase their risk of harm. How our data collection fits into a mosaic comprised of many sources, some of which might not be collected yet, thus poses an unusually thorny ethical problem.

Imagine, for example, fielding a survey that records the respondent's ethnic or religious identity. With standard random samples, a site's ethnic composition could be extracted with reasonable precision simply from the respondents' demographic information. Standard data practices, including redacting respondents' names, would not guard against the construction of these site-level indicators. In turn, these demographic data, including questions about education and wealth, could be married with data gleaned from other methods (e.g., remote sensing) to assemble a rich mosaic of site-specific attributes without ever compromising informed consent or respondent anonymity. Repressive governments could use these mosaics to target rival ethnic groups or pro-insurgent populations. Humanitarian organizations might use these data portraits to direct their aid efforts. A researcher might find themself inadvertently helping choose winners and losers, or fueling conflict dynamics, if their DII data are combined with other efforts.

Such data collection efforts are likely to be clandestine and, as such, may be hard to detect. Evidence suggests, however, that they are already underway. In 2017–2018, for example, the Bangladeshi government joined forces with the UNHCR and various humanitarian aid organizations to register nearly 830,000 Rohingya refugees who had fled genocidal violence in neighboring Myanmar. Surveys, "smart cards," and other biometric data, including fingerprints and photographs, were merged with cadastral mapping from drones to create a composite "mosaic" record for every single refugee, family, and household across dozens of camps. These records were then used to identify and forcibly repatriate more than 20,000 refugees back to Myanmar, despite widespread protests. A follow-up investigation by Human Rights Watch (HRW) determined that many refugees did not provide informed consent for the collection, let alone actual use, of these personalized data (HRW 2021a). Nonetheless, in the face of severe overcrowding at Cox's Bazar, the authorities once again used these records to select large families for forced resettlement to Bhasan Char, a dismal island where dozens have drowned trying to escape back to the mainland (HRW 2021b).

Scholars working in Cox's Bazar rely on the same data architecture—including UNHCR-provided camp and subcamp locations—established by these earlier government efforts. As a result, their publicly available survey data, even if scrubbed of personal information, can still be merged with prior UNHRC data collection efforts. Bangladeshi police, notorious for their corruption, could use survey data to refine their bribery and extortion efforts by mapping relative concentrations of wealth within the camps. These surveys also routinely collect fine-grained data on violent incidents that, when paired with publicly available geocoded data on protests, could be used to justify additional crackdowns on movement within the camps (HRW 2022; Al Jazeera 2022).

The need to consider future data uses is only likely to grow. Cell phone records, for example, are often specific enough to identify individuals through point purchasing patterns even if names and phone numbers are removed. Standard wealth indicators ("how many live in your home?"; "what is your roof made of?") can be cross-referenced with satellite imagery to identify specific interviewees or survey respondents. Even social media or photographs taken from field sites can be investigated using open-source techniques to identify locations and individuals, potentially increasing the risk of discovery and possible harm. We are fast closing the loop between inadvertent exposure of research subjects and groups and the prospect that such lapses can be exploited by (bad) actors in near real time.

Note, too, that the convergence of these data streams raises a morass of tricky issues for prevailing notions of informed consent. Merging data collected under informed consent with other data where consent was neither asked for nor given (e.g., via remote imagery) may invalidate the original consent. Individuals might also give consent while unaware that their data will be merged with other sources. Our preregistration documents should therefore consider explicitly the consequences for groups affected by the proposed research. This amounts to a major revision of how current IRB practices are structured and how we view ethical fieldwork.

Central to this radical reimagining will be extending our ethical conversations to adjacent communities. Certain interventions (e.g., hot-spot policing) can displace crime from targeted communities to neighboring ones. Aid, too, can draw new populations to the disbursement site or, alternatively, create jealousy among neighboring villages that do not receive aid. Paying locals to play experimental behavioral games might similarly create anger or resentment among excluded neighbors. Kaplan's policy engagement with a community member possibly created jealousies among other individuals who were not invited or unable to participate (Kaplan, Chapter 5, this volume). Even focus groups and interviews could be misinterpreted as a prelude to future aid (or government presence), shifting expectations and behavior in unvisited adjacent communities.[26] How far the effects of our research might ripple is a matter of the nature of the intervention and the local context. But the notion that such spillover is possible—indeed, likely—should inform our preregistration exercise.

Policymakers and the Policy Process

There is little formal guidance to date on what constitutes "responsible" engagement of policymakers or, indeed, the broader public. Yet, the nature of our involvement in the policy process and the end uses of our research have far-reaching implications for vulnerable populations and home audiences alike. While each of us will define our role differently, researchers more broadly must consider their desired position vis-à-vis the subjects of their study and in the policy process before fieldwork begins (see Kaplan, Chapter 5, this volume). We cannot be stewards of ethical policy advice if our own research is ethically flawed at the outset. But we must work to maintain our ethical redlines as we become enmeshed in the policy process and dissemination of findings.

Researchers seeking to embed themselves within (non)governmental organizations will need to introduce these partners to their ethical redlines very early in any collaboration. Sharing a preregistration document can ensure that ethical concerns are reflected in the choice of research design, nature of the proposed intervention and fieldwork, and post-fieldwork public outreach. Key questions to consider include: What is the level of agreement about ethical redlines and proposed risk and harm mitigation strategies between the lead researcher and partners? Does dependence on a research partner create vulnerabilities that might lead to e-hacking? Is one willing to consent to prepublication review and accept limitations on data-sharing? Will the research

[26] Spillover also raises the question of how we might compensate adjacent individuals and communities for harmful effects when they are not formally part of our fieldwork.

be peer reviewed before offering actionable policy advice? Is there agreement on public outreach? Are side analyses allowed that will be used for internal organizational purposes but not published? Can (initial) findings being injected directly into the policy process?

Even indirect routes to influencing the policy process can raise ethical issues. We can now reach broad audiences through high-quality e-magazines and blogs (e.g., *The Washington Post*'s "Monkey Cage", and its successor, "Good Authority", *Political Violence at a Glance*, *Foreign Affairs*, and *War on the Rocks*, podcasts, op-eds, and social media like Twitter/X, BlueSky, LinkedIn, and Meta). Doing so, however, can sometimes force a trade-off between nuance and flashiness. Weighed down by caveats, our arguments may fail to find traction among harried policymakers looking for streamlined solutions (see Sisk, Chapter 4, this volume; Steadman 2023). Whether we should junk nuance for simpler narratives, and how we convey uncertainty about our findings, will remain challenges for policy-facing scholars (see Epstein, Chapter 7, this volume). We will also wrestle with the moral hazard problem of supporting certain policies or interventions where we are insulated from direct consequences of failure (Chenoweth 2020). Public efforts to shape policy might also meet unwelcome resistance if the policy issue has become politicized.[27] Ultimately, we cannot control how policymakers or the public engage our work. But we can pre-register our commitment to how we describe our findings, what types of outlets and audiences we will engage, and our level of engagement in fact-checking and challenging mistaken narratives.

Protecting Redlines: Specific Actions

Staking out ethical redlines is only half the battle, however. Ethics must be translated into action if we are to conduct fieldwork and policy engagement responsibly. I therefore offer eight suggestions that, when taken together, provide a blueprint for actionable items that could be included in a preregistration document.

1. *Formalize your risk assessment.* All preregistration documents should provide a transparent accounting of the possible risks associated with the proposed fieldwork. While context-specific, risk assessments should include a detailed discussion of the baseline risk to the vulnerable

[27] Online harassment and abuse is especially severe for women and people of color. While one in five adults report being harassed online for their political views, women and POC are more likely to experience severe harassment and to be targeted for their gender or ethnicity. See Vogels 2021: 8–11.

population as well as specific research partners, communities, and adjacent populations. Some research may require a statement of how the proposed intervention will affect both recipient and adjacent communities and locations. Risk assessment should explicitly acknowledge the possibility that risk may spill over to neighboring populations. Once a baseline is established, researchers should also outline what types of actions represent increased risk to research participants and facilitators. The threat of increased harm might arise from the research itself or from changing circumstances in the fieldwork setting. In either case, researchers should detail how much increased risk is acceptable before the net balance between the costs and benefits of research swings too sharply toward anticipated harm.[28]

2. *Define informed consent.* Scholars should detail how informed consent will be obtained from both research participants and host communities. While consent is contextually specific, researchers must nonetheless describe how participants will be informed about the potential risks of participation, including the level of uncertainty about those risks (Wood 2006). Scholars should pay attention to both the quality of informed consent and the types of information that would enable participants to make credible assessments about the costs and benefits of participation. How permission will be obtained, and from which consenting parties, must also be described. In many settings, informed consent is a combination of individual decisions and those of relevant political authorities, including village elders, regional administrators, and neighborhood organizations.

Consent is also an ongoing process rather than a one-time event. Researchers could pose follow-up questions during the research process to ensure that participants still wish to participate. These questions could also probe for issues or practices that are causing unease. In one recent field experiment in Mexico, for example, researchers pre-specified a stopping rule: if 5 percent or more of participants felt unhappy about their participation, the study would pause so that additional information could be given to assuage their concerns (Baron et al. 2022). At the extreme, researchers might redesign the entire project and its consent process if the threshold of concerned participants is breached.

3. *Sign a memorandum of understanding (MOU) with relevant partners.* MOUs can be a powerful vehicle for safeguarding ethical redlines.

[28] Examples of specific questions to pose when assessing risk can be found in Asiedu et al. 2021.

Researchers should craft and sign these documents with relevant partners well before conducting fieldwork or disseminating findings. MOUs should determine who controls the data that results from the partnership; whether partner organizations have the right to prepublication review; whether the researcher will be expected to conduct "off-book" analyses intended for internal consumption only; and the nature of post-research media outreach efforts, including the extent of the researcher's participation in these activities. Detailing responsibilities clearly, as well as timelines for deliverables, can also alleviate time constraints that often weaken ethical safeguards. While legalistic, these MOUs can be invaluable for protecting a researcher's intellectual freedom and for shaping how findings are portrayed to policy audiences and the broader public.

4. *Commit to robust piloting.* We often operate without knowing fully how the proposed research will affect local populations. This is especially true of large-scale field experiments and surveys on sensitive issues in (post)conflict settings. Even focus groups and interviews have the potential for creating unexpected harm if they upset local dynamics. Researchers should therefore plan (and budget) for piloting their interventions, surveys, or interview questions on a relevant subset of the population before scaling up their efforts. Desk reviews of prior research are of course useful. But little can replace the lessons learned from piloting in the same conditions where the proposed research will be conducted. Such efforts might reveal that certain questions are too sensitive, that unexpected dynamics arise in focus groups, or that substantial spillover effects are present, forcing adjustments to either the intervention or the measurement strategy. Pilots should be large enough to detect potential problems; multiple rounds might be necessary to fine-tune risk assessments and to avoid inflicting harm inadvertently.
5. *Build a "kill-switch."* Based on their risk assessment, researchers should build an explicit mechanism—a "kill-switch"—for rapidly suspending fieldwork. All researchers, regardless of their methodological leanings, should be able to identify a set of indicators for determining whether their research is inadvertently elevating risk or causing actual physical or mental harm. Surges in crime or violence relative to pre-fieldwork baselines offer one possible warning sign. Adverse behavioral changes among participants are also important clues. Perhaps, for example, a police training program makes officers more abusive toward civilians; perhaps cash transfers stoke increased domestic violence by upending traditional gender norms. Within-subject observations, especially

through interviews and surveys, will be especially useful in detecting increased risk or harm.

Timing is key: investigators should engage in persistent monitoring over the project's life, often at repeated intervals, rather than a one-time evaluation at the project's end. Multiple attitudinal and behavioral indicators could be blended into an index for assessing when to pause fieldwork. Of course, not every adverse incident will require pausing fieldwork. Instead, researchers will need to draw conclusions about whether the observed harm stems from the researcher's fieldwork or from changing ground conditions unrelated to the researcher's presence. Criteria for restarting fieldwork should also be discussed in the preregistration document.

6. *Consider debriefing and post-research harm mitigation.* Scholars should consider whether participants and relevant communities should be debriefed once fieldwork ends. Interviews and focus groups could be conducted with participants and key local stakeholders to explain the fieldwork's purpose and to gauge downstream harm.[29] Participants could also be asked if they believed the research was ethical and appropriate, particularly if deception is used (Desposato 2018). Public meetings to describe the research and its findings would also represent a good-faith effort to give back to host communities while also investigating the possibility of post-research harm (see Kaplan, Chapter 5, this volume). These meetings are especially valuable after fieldwork that involves attitudinal or behavioral interventions that seek to alter attitudes toward out-groups or coping strategies among vulnerable populations. They also represent a relatively low-cost measure of collecting additional data on the spatial reach and temporal persistence of our interventions.

Scholars should pair debriefing with pre-registered plans for mitigating possible second-order consequences that might arise for host or adjacent communities. To avoid "informational harm," scholars should redact community-identifying information, including latitude and longitude coordinates, from public datasets (Raymond 2019).[30] Researchers, along with their partners, can also establish data collection plans to monitor relevant communities for post-research harm. While we typically view research as concluded once we

[29] See, for example, Lyall, Zhou, and Imai 2020.
[30] These could be made available to other scholars for replication or publication purposes so long as community-level data remained anonymized.

leave the field, the reality is that the effects of our interventions (and even just our presence) can take time to unfold. In some cases, the negative consequences of these interventions may manifest only after prying (foreign) eyes have left. Above all, scholars should use the preregistration exercise to consider what obligations they owe to participants and host communities and how long these debts last.

7. *Identify possible misuse of research by policymakers.* One downside of policy engagement is the potential that our findings and, in some cases, even our methodological tools, might be applied for purposes with which we may disagree on ethical grounds (see DeMartino, Chapter 6, this volume).[31] At worst, our findings might be miscast to defend odious policies or inflict actual harm on vulnerable populations. To be clear, scholars are not responsible for every misuse of their data or findings. Nor can we let policymakers exercise a kind of soft veto over the types of research that we conduct.

Instead, scholars should consider foreseeable misuses of findings and act to mitigate these concerns. For example, concealing village locations that receive "hearts and minds" economic assistance would prevent governments from punishing communities that continued to support criminal or insurgent organizations.[32] How findings are disseminated is also important. Social media platforms like The Washington Post's "Monkey Cage" powerful channels for influencing policy. But outreach can sometimes come at the expense of nuance and exaggerated findings (see Epstein, Chapter 7, this volume). Scholars should consider how they might effectively communicate both the strength of the findings and possible caveats, including levels of uncertainty, before a media blitz is launched.[33] This outreach should extend to the donors that funded the research.

It is also possible to build safeguards around your ethical redlines. Let me provide a brief example from my own experience. As part of a Defense Department Minerva grant, I was provided with declassified

[31] Consider, for example, the possibility that the creation of statistical packages or the automation of text analysis provides actors with new technical capabilities for using (or extending) your original analysis.

[32] As part of USAID's MISTI initiative, I was tasked with measuring support for the Afghan government using very large public opinion surveys (2012–2015). After an initial round revealed low public support for the Ghani administration, USAID prohibited the sharing of subsequent reports and data to prevent possible retaliation against "disloyal" populations.

[33] We also lack a norm about whether our research should be peer reviewed (or, despite prolonged lag times, even published) before publicizing our findings. The average "Monkey Cage" post, for example, earns tens of thousands of unique views—far more than the vanishingly small number of scholars who will engage our work behind paywalls. Yet these outlets have no mechanism for verifying our findings or correcting them if mistakes are discovered.

airstrike data from Afghanistan and Iraq. The purpose of the project was to assess the effects of airpower on insurgent violence and civilian attitudes down to the village level (Lyall 2019; Papadogeorgou et al. 2022). The results were dismal: airstrikes were associated with increases in the frequency and lethality of insurgent attacks in and around the bombed locations.

I then began presenting my findings to several high-level audiences both in the US and overseas. After one such presentation, I was approached by a group of individuals who were unexpectedly excited by my findings. Why? The insurgent responses were so predictable in location and timing that they could be used to anticipate insurgent behavior, helping the Air Force sharpen its ability to target them. This could be hugely valuable on the battlefield, they noted. Would I be willing to run my analyses in near-real time to facilitate targeting sorties, in exchange for expanded access to (near) current data? I declined: I wanted to understand the violence, not engage in its actual production. I therefore imposed a six-month delay in receiving new data updates, a form of planned obsolescence of my findings that prevented them from being used to guide daily targeting decisions.

8. *Know when to exit.* While optimism is a prerequisite of fieldwork in fragile settings, researchers must also consider worst-case scenarios. Above all, scholars should be prepared to answer the question of when (and how) to shutter a project completely if ethical redlines are breached. Some situations may simply prove irredeemable from an ethical standpoint. These conversations must extend to both collaborators and the donor organizations funding the research.

Committing to an exit strategy in advance can prevent situations where researchers find themselves rationalizing ethical violations once they are enmeshed in the hurly-burly of fieldwork. It is also important to recognize that participants, brokers, and host communities do not have the same exit options. They will bear the brunt of ethical transgressions and the consequences of hasty exits. Before opening the door to fieldwork, it makes sense to locate the exits and anticipate the circumstances that might drive you to them.[34]

[34] This logic also applies to collaboration with policymakers in which one's findings or methods are distorted for political purposes. An exit—perhaps as a public correction (or disavowal) of the research, or a private severing of ties—may be necessary under certain conditions.

Better Social Science and Policy through Preregistration

There is little use hiding the fact that preregistration, if done correctly, is burdensome. It represents a substantial time commitment and a costly, if self-imposed, constraint on a researcher's discretion during fieldwork. Yet the advantages for both social science and policy are considerable. At a minimum, preregistration is a boon to transparency: often-hidden decisions about risk assessment, harm mitigation, and methodological choices are now revealed in far greater detail than current norms dictate.[35] Posting these documents, whether as part of a formal preregistration exercise or a more personal statement, will also help build a repository of best practices about how to approach ethical fieldwork and policy engagement. Indeed, in the absence of a universal code of conduct for political scientists,[36] these statements may collectively represent both an emergent norm and a learning tool for scholars, especially those approaching fieldwork for the first time.

Reputational benefits and, yes, costs, might further cement this norm into place. As more researchers pre-register their redlines, organizational partners might choose to work only with scholars who have publicly signaled their commitment to ethical research. By contrast, unscrupulous organizations might steer clear of researchers who have signaled their ethical commitment, reducing the odds that scholars will confront ethical dilemmas once in the field. And, on the methodological front, the need to assess risk and monitor possible harms cries out for a mixed-methods approach, with qualitative and quantitative approaches joined in a common framework. Interviews and focus groups, for example, can be conducted alongside a field experiment to determine why a particular result was observed and whether harm was inflicted on research participants. The result is more rigorous, and perhaps more nuanced: findings can improve the design and implementation of policy interventions in these settings.

A note of caution is warranted here, however. The radical transparency offered by preregistration has the potential to create more conflict among scholars, at least initially. Nothing comes for free; in this case, transparency about where we choose to draw our redlines will dig up subterranean disagreements that we currently bury in the anonymous peer review process

[35] The situation is slowly changing, however. Scholars have started calling for ethical discussions in appendices (Asiedu et al. 2021), while leading journals (notably, *American Political Science Review*) have mandated ethics statements.

[36] For one such effort in economics, see DeMartino 2011.

in the hopes that it will quietly screen out unethical actors. Debates over the location of our redlines might arise both within academic teams and across methodological divides. Past unethical conduct, or onerous restrictions on data access, might lead to certain organizations and companies findings themselves blacklisted by some potential collaborators. Reputational harm might accrue to scholars who continue working with such organizations, creating conflict over where to draw ethical boundaries and who gets to police them. Transparency, in other words, might make for better, but not necessarily more harmonious, social science. Yet this state of affairs is preferable to the status quo, where ethical disagreements are rarely broached openly.

As a final note, the demands of preregistration should not become so onerous that they crowd out fieldwork entirely. Preregistration is a framework, not a straitjacket. Scholars should also take care to revisit their preregistration documents after major rounds of fieldwork or before a new project is undertaken. Past ethical concerns, for example, may be irrelevant in a new context. Precautions about collecting certain types of data or asking particular questions might be relaxed if individuals prove far more forthcoming, and the risks of harm much lower, than anticipated. The opposite might also hold true: with experience, researchers might shift their ethical redlines forward as they come to better realize the sources of risk and harm in a given setting. Preregistration documents are, in the final analysis, living documents that should reflect our hard-won lessons from engaging policymakers and conducting fieldwork.

Conclusion

This chapter has sketched a modest proposal: we should publicly pre-register our ethical redlines when conducting policy-relevant fieldwork among vulnerable populations. While a simple practice, preregistration has numerous benefits. These include increased transparency of risk assessments, stronger safeguards against unexpected harm to research participants and communities, and improved communication of study limitations to policymakers. We need, in other words, to move beyond the narrow, often legalistic, definitions of harm and risk that are embedded in current IRB practices. Doing so will generate benefits that accrue regardless of a scholar's methodology: interviews, focus groups, participant observation, surveys, and randomized control trials could all be housed within a preregistration framework.

To be clear, my intent here is not to impose a single ethical standard. Many issues that arise during fieldwork are context-specific and morally challenging; individuals acting in good faith will reach different conclusions about how to address ethical dilemmas. But all scholars, their partners, and funding organizations will benefit from investing time and effort identifying ethical redlines, assessing baseline risks, and proposing possible solutions before fieldwork. The stakes, already high when dealing with vulnerable populations, only become more so when we step into the policy arena, regardless of our good intentions. Public pledges of our redlines can help ensure that we remain within the guardrails of ethical fieldwork and policy engagement while avoiding the pitfalls of our ambitions.

References

AAA Commission on the Engagement of Anthropology with the US Security and Intelligence Communities (CEAUSSIC) (2009) Final Report on the Army's Human Terrain System Proof of Concept Program. Technical report.

Al Jazeera (2022) "'It's hell': Rohingya Refugees in Bangladesh Rally to 'Go Home.'" June 19. https://www.aljazeera.com/news/2022/6/19/rohingya-refugees-in-bangladesh-rally-to-go-home.

American Political Science Association (2020) "Principles and Guidance for Human Subjects Research." https://connect.apsanet.org/hsr/principles-and-guidance/.

Asiedu, Edward, Dean Karlan, Monica Lambon-Quayefio, and Christopher Udry (2021) "A Call for Structured Ethics Appendices in Social Science Papers." Proceedings of the National Academy of Sciences 118(29): 1–10.

Baaz, Maria and Mats Utas (2019) "Exploring the Backstage: Methodological and Ethical Issues Surrounding the Role of Research Brokers in Insecure Zones." *Civil Wars* 21(2): 157–178.

Baron, Hannah and Lauren Young (2021) "From Principles to Practice: Methods to Increase the Transparency of Research Ethics in Violent Contexts." *Political Science Research and Methods* 10(4): 840–847.

Baron, Hannah, Omar Garcia-Ponze, Lauren Young, and Thomas Zeitzoff (2022) *The Limits of Deliberation: A Field Experiment on Criminal Justice Preferences in Mexico*. Unpublished Manuscript.

Blair, Graeme, Jasper Cooper, Alexander Coppock, and Macartan Humphreys (2019) "Declaring and Diagnosing Research Designs." *American Political Science Review* 113(3): 838–859.

Blair, Robert and Michael Weintraub (2021) *Military Policing Exacerbates Crime and May Increase Human Rights Abuses: A Randomized Controlled Trial in Cali, Colombia*. Unpublished Manuscript.

Bursztyn, Leonardo, Davide Cantoni, David Yang, Noam Yuchtman, and Y. Jane Zhang (2021) "Persistent Political Engagement: Social Interactions and the Dynamics of Protest Movements." *American Economic Review: Insights* 3(2): 233–250.

Campbell, Susanna (2017) "Ethics of Research in Conflict Environments." *Journal of Global Security Studies* 2(1): 89–101.

Cheeseman, Nic and Caryn Peiffer (2022) "The Curse of Good Intentions: Why Anticorruption Messaging Can Encourage Bribery." *American Political Science Review* 116(3): 1081–1095.

Chenoweth, Erica (2020) "On Research that 'Matters.'" In *Stories From the Field: A Guide to Navigating Fieldwork in Political Science*, ed. Peter Krause and Ora Szekely. New York: Columbia University Press, pp. 267–276.

Cilliers, Jacobus, Oeindrilla Dube, and Bilal Siddiqi (2015) "The White-Man Effect: How Foreigner Presence Affects Behavior in Experiments." *Journal of Economic Behavior Organization* 118: 397–414.

Cronin-Furman, Kate and Milli Lake (2018) "Ethics Abroad: Fieldwork in Fragile and Violent Contexts." *PS: Political Science & Politics* 51(3): 607–614.

DeMartino, George (2011) *The Economist's Oath: On the Need for and Content of Professional Economics Ethics*. Oxford: Oxford University Press.

Desposato, Scott (2018) "Subjects and Scholars' Views on the Ethics of Political Science Field Experiments." *Perspectives on Politics* 16(3): 739–750.

Devereaux, George (1967) *From Anxiety to Method in the Behavioral Sciences*. The Hague: Mouton.

Diao, Mi, Hui Kong, and Jinhua Zhao (2021) "Impacts of Transportation Network Companies on Urban Mobility." *Nature Sustainability* 4: 494–500.

Fearon, James (1997) "Signaling in Foreign Policy Interests: Tying Hands Versus Sinking Costs." *Journal of Conflict Resolution* 1: 68–90.

Fujii, Lee Ann (2012) "Research Ethics 101: Dilemmas and Responsibilities." *PS: Political Science & Politics* 45(4): 717–723.

Hoover Green, Amelia and Dara Cohen (2021) "Centering Human Subjects: The Ethics of 'Desk Research' on Political Violence." *Journal of Global Security Studies* 6(2): ogaa029.

Human Rights Watch (2021a) "UN Shared Rohingya Data Without Informed Consent." https://www.hrw.org/news/2021/06/15/un-shared-rohingya-data-without-informed-consent.

Human Rights Watch (2021b) "Bangladesh: Halt Forced Relocation of Rohingya Refugees." https://www.hrw.org/news/2021/11/24/bangladesh-halt-forced-relocation-rohingya-refugees.

Human Rights Watch (2022) "Bangladesh: New Restrictions on Rohingya Camps." https://www.hrw.org/news/2022/04/04/bangladesh-new-restrictions-rohingya-camps.

Humphreys, Macartan, Raul Sanchez de la Sierra, and Peter van der Windt (2013) "Fishing, Commitment, and Communication: A Proposal for Comprehensive Nonbinding Research Registration." *Political Analysis* 21(1): 1–20.

Loy, Irwi (2021) "Biometric Data and the Taliban: What are the Risks?" *The New Humanitarian*, September 2. https://www.thenewhumanitarian.org/interview/2021/2/9/the-risks-of-biometric-data-and-the-taliban.

Lyall, Jason (2019) *Bombing to Lose? Airpower and the Dynamics of Coercion in Counterinsurgency Wars.* Unpublished Manuscript.

Lyall, Jason, Yang-Yang Zhou, and Kosuke Imai (2020) "Can Economic Assistance Shape Combatant Support in Wartime? Experimental Evidence from Afghanistan." *American Political Science Review* 114(1): 126–143.

McDermott, Rose and Peter Hatemi (2020) "Ethics in Field Experimentation: A Call to Establish New Standards to Protect the Public From Unwanted Manipulation and Real Harms." *Proceedings of the National Academy of Sciences* 117(48): 30014–30021.

Nair, Gautam and Nicholas Sambanis (2019) "Violence Exposure and Ethnic Identification: Evidence from Kashmir." *International Organization* 73: 329–363.

National Committee for the Protection of Human Subjects of Biomedical and Behavioral Research (1979) *The Belmont Report: Ethical Principles and Guidelines for the Protection of Human Subjects of Research.* Washington, DC: Government Printing Office.

Nordstrom, Carolyn and Antonius C. G. M. Robben, eds. (1996) *Fieldwork Under Fire: Contemporary Studies of Violence and Culture.* Berkeley: University of California Press.

Nosek, Brian, Charles Ebersole, Alexander DeHaven, and David Mellor (2018) "The Preregistration Revolution." *Proceedings of the National Academy of Sciences* 115(11): 2600–2606.

Oliver, Kathryn and Paul Cairney (2019) "The Dos and Don'ts of Influencing Policy: A Systematic Review of Advice to Academics." *Palgrave Communications* 5: 1–11.

Page, Scott (2017) *The Diversity Bonus: How Great Teams Pay Off in the Knowledge Economy.* Princeton, NJ: Princeton University Press.

Papadogeorgou, Georgia, Kosuke Imai, Jason Lyall, and Fan Li (2022) "Causal Inference with Spatio-temporal Data: Estimating the Effects of Airstrikes on Insurgent Violence in Iraq." *Journal of the Royal Statistical Society (Series B)* 84(5): 1969–1999.

Raymond, Nathaniel (2017) "Beyond "Do No Harm" and Individual Consent: Reckoning with the Emerging Ethical Challenges of Civil Society's Use of Data." In *Group Privacy: New Challenges of Data Technologies*, ed. Linnet Taylor, Luciano Floridi, and Bart van der Sloot. New York: Springer, pp. 67–82.

Raymond, Nathaniel (2019) "Reboot Ethical Review of the Age of Big Data." *Nature* 568: 277.

Vogels, Emily (2021) *The State of Online Harassment*. Washington, DC: Pew Research Center.

Wood, Elisabeth (2006) "The Ethical Challenges of Field Research in Conflict Zones." *Qualitative Sociology* 29(3): 373–386.

Zingales, Luigi (2014) "Preventing Economists' Capture." In *Preventing Regulatory Capture*, ed. Daniel Carpenter and David Moss. Cambridge: Cambridge University Press, pp. 124–151.

3
Sharing Unpopular Findings
When Academic Research Challenges Policy Efforts

Tricia D. Olsen

What are the challenges of sharing unpopular findings with policymakers? In 2012, a coauthor, Professor Leigh Payne of the University of Oxford, and I were contacted by a member of the UN Working Group (UNWG) on Business and Human Rights. They reached out to inquire whether we might gather systematic data on businesses and their human rights records. The UNWG was interested in data-driven policymaking and the possibilities of how data might apply to their current work informed by the UN Guiding Principles on Business and Human Rights. The UNWG member was familiar with previous databases my coauthors and I had created (Olsen, et al. 2010; Dancy, et al. 2019), and thus suggested that we might create a database on corporate human rights abuses to "show that the UN Guiding Principles work." We quickly clarified what research and data could accomplish, stating that while we would collect these data (which, today, is the Corporations and Human Rights Database-Latin America, or CHRD-LA[1]), the results may not necessarily show that the United Nations Guiding Principles on Business and Human Rights (hereafter, UNGPs) shape corporate behavior or respect for human rights. We continued with a classic (though true) academic retort: "That," we said, "is an empirical question" and one we would be very interested in exploring. My colleague and I were under the impression that we had clarified any misconceptions and so the conversation continued. Yet, in retrospect, this off-hand comment embodied the challenges and lessons that lay ahead.

Data collection efforts often inform policymakers and can greatly contribute to ensuring policies are effective and appropriate. At the time we were contacted by the UNWG member, there was no comprehensive data about allegations of corporate human rights abuses. Moreover, there had been no efforts to collect systematic information about when victims have

[1] For additional information, please see www.bhrlab.com.

Tricia D. Olsen, *Sharing Unpopular Findings*. In: *Speaking Science to Power*. Edited by: Rachel A. Epstein and Oliver Kaplan, Oxford University Press. © Tricia D. Olsen (2024). DOI: 10.1093/oso/9780198875185.003.0003

access to judicial or nonjudicial remedy. I viewed this as a unique opportunity to collect data that would be welcomed by, and inform, the policymaking community. After all, the UNWG had expressed their interest in these data, were encouraging complementary data collection efforts, and had expressed interest in a thoughtful approach to data-driven policymaking. It seemed like an academic's dream; an opportunity to make a meaningful contribution to practice and scholarship. While we embarked on the data collection efforts, which have resulted in numerous scholarly contributions that I (still) hope might inform policy (Payne et al. 2020; Olsen et al. 2022; Olsen 2023; Olsen and Bernal-Bermúdez 2023), the path to doing so was less straightforward than I initially imagined. A central dilemma ultimately arose: the data suggested that policies already being implemented by the UNWG were perhaps not the right policies.

This case study begins by outlining how the engagement moved forward after our initial conversation and, later, how I shared findings at UN meetings, not having realized that many of those involved were not open to new information at that time. I then consider how false assumptions and misaligned incentives shaped the engagement. The case study ends by reflecting on possible outcomes stemming from the sharing of unpopular findings and explores two questions at the heart of this engagement: How does one criticize or challenge an institution and its stakeholders when their good will is necessary for bringing about better policy outcomes? And, second, when sharing unpopular findings, what obligations, if any, do scholars have when policymakers do not care to hear the message?

Engagement from a Distance

The initial impetus for our research project began with our conversation with the UNWG, but we were not engaging in the research *with* the UNWG. It may have been because of the normative comment in which our UNWG contact hoped for results that supported their priors, but we began to collect data independently to ensure the academic integrity of our work.[2] We maintained contact with our UNWG partner (indeed, I was an external consultant for related survey research for the UNWG) but did not closely collaborate on the Corporations and Human Rights Database-Latin America. We found the topic we were charged with to be intellectually intriguing given some of the

[2] While the UNWG was interested in data about business and human rights at the time and encouraged this work, there was no UN-based funding to support the creation of the Corporations and Human Rights Database Project.

persistent debates in the literature and the dearth of data on corporate human rights abuses and remedies for wrongdoing (Olsen 2023).

With enthusiasm, we quickly got to work and sought funding for a pilot project. We secured some seed money to support the research and began collecting data on allegations of corporate human rights abuses in Latin America.[3] The data mirror the structure of the UNGPs, which the UN Human Rights Council unanimously endorsed in June 2011 (Figure 3.1). The three pillars of the UNGPs are: 1) the state duty to protect human rights; 2) the corporate responsibility to respect human rights; and 3) that all actors should work to help victims access remedy mechanisms when abuses do occur.

Figure 3.1 Tricia Olsen attended the UN Working Group on Business and Human Rights meeting at the Human Rights Council in Geneva, Switzerland, 2015. The Council adopted a resolution in 2011 creating the Working Group to "promote the effective and comprehensive dissemination and implementation" of the Guiding Principles on Business and Human Rights.

Photo: Tricia Olsen

[3] I am grateful for funding from the University of Denver's PROF grant and Internationalization Grant; Payne and I also secured seed funding from the British Academy-Leverhulme and the University of Oxford Press John Fell Fund for their early support of this project. In 2019, I received funding to create the global dataset from the National Science Foundation (Award Number #1921229); see www.bhrlab.com.

The CHRD is the first database of its kind that offers a tool for analyzing five types of allegations of corporate human rights abuse (physical violence, development and poverty, health, environment, and labor) and judicial or nonjudicial remedy mechanisms associated with each allegation. While today's CHRD is global, with data on more than 5,400 allegations of corporate human rights abuse, the CHRD-LA pilot project only included data on Latin America. The CHRD-LA is comprised of allegations of corporate involvement in human rights abuses across the Latin American region. The main source of that data is an online archive of claims spanning from 2000 to 2014 hosted by the Business and Human Rights Resource Centre (BHRRC). The database offers a full range of business sectors, firms, and country contexts. The CHRD-LA includes multiple levels of data: the emergence of claims, their institutionalization in formal and informal remedial processes, and their outcomes, which include judicial and nonjudicial remedy efforts. The CHRD-LA offers a large-*N* comparative approach that includes more than 1,300 allegations against 916 firms, with important variation on firm size, headquarter location, ownership structure, types of abuse, the profile of victims of these abuses, and the evolution and outcome of claims processes.

With a completed dataset for Latin America and preliminary findings in hand, my UNWG contact began seeking out opportunities for me to share our work with the remaining UNWG members and other key stakeholders. I was invited to present the research at two different UN-sponsored events. I learned several key lessons from those engagements, not least of which was how to overcome some complex engagement challenges, which I explore in the following sections.

Challenge 1: Clarifying Your Role as a Scholar

As the anecdote at the beginning of the case illustrates, there was a lack of clarity as to the role that research would play in this particular context. Our contact at the UN Working Group assumed that our research would help achieve their policy goals. Meanwhile, we thought our quick clarification about what the data may, or may not, show had assuaged any misconceptions about our role as scholars. In retrospect, at that juncture we should have embarked on a more detailed conversation that would have helped us grasp the UNWG member's position, not only within the broader structure of the UN but also with respect to how the Working Group viewed their role vis-à-vis external stakeholders (e.g., human rights advocacy groups, organizations critical of the UNGPs, and corporate partners). At the time, I did not appreciate the challenges associated with sharing data with policymakers, nor the degree to which our interests were (mis)aligned with those of the UNWG. Despite our initial, though brief, clarification that the evidence from our research could yield positive *or* negative conclusions about the effectiveness of the UN Guiding Principles, I realized later, while I was sharing the research, that members of the UNWG continued to assume our results would be supportive of their agenda.

I did not anticipate this discrepancy and expected that my engagement—regardless of the results—would be welcomed and understood. I assumed this, in part, because of the weighty role scholars played, historically, in the policy debate about whether hard law (e.g., formal laws with penalties) or soft law (e.g., voluntary norms) serve as better tools to improve corporate respect for human rights. Those advocating for hard law suggest that individual rights will only be respected when the rule of law is equally applied to all (Humphrey 2003: 498; see also Méndez 1997; Osiel 2000). Moreover, many liberal rule of law advocates seek to address the historic absence of accountability for corporate actors across both developed and developing countries by promoting international customary law. Scholars and policymakers suggested that states were either unable or unwilling to hold corporate actors accountable and that a global legal solution was needed to fill this governance gap.

In the 1990s, the UN gathered a group of five independent experts—spearheaded by Professor David Weissbrodt of the University of Minnesota—to draft a code of conduct for transnational corporations, or what is now known as the "Norms on the Responsibilities of Transnational Corporations and Other Business Enterprises with Regard to Human Rights" (hereafter the "Draft Norms"). The Draft Norms set out human rights standards for companies and drew heavily from existing international law in areas such as civil, political, economic, social, and cultural rights that addressed issues ranging from labor concerns to environmental issues to consumer protection (for a more detailed account, see Mantilla 2009).

While years of work went into creating the Draft Norms, the effort was abruptly discarded in 2004 due to a lack of support from key state and corporate actors. In the wake of that disagreement, many within the United Nations were concerned that the business and human rights agenda could have been completely abandoned. To revive the discussion, UN Secretary-General Kofi Annan appointed Harvard University Professor John Ruggie as the Special Representative for Business and Human Rights. Ruggie was selected, in part, because of his success in generating support for The Global Compact, which was launched in 2000.

The Global Compact is a voluntary, soft law mechanism that facilitates discussions and best practices around corporate behavior and human rights through ten key principles. It urges companies to support and respect the protection of internationally recognized human rights and to ensure, through voluntary reporting, that they are not complicit in the abuse of those rights. The Global Compact promotes "shared values and principles" for state and nonstate actors alike and represents one of the first efforts to encourage nonstate actors to publicly commit to respecting human rights. As of 2024, the Global Compact had 9,841 corporate members (of which 9,565 are considered active) and was the precursor for today's leading business and human rights initiative: the UNGPs. Ruggie adopted an approach of "principled pragmatism," which is a combined effort of voluntary compliance and the promotion of global norms—an approach that draws from his academic work (Ruggie 2013). The debate between hard and soft law continues today, however, as the draft treaty on business and human rights has involved the UNWG in ongoing negotiations.

This background matters because it highlights the prominent role scholars played in shaping policy at the UN. When I first began to engage with the UNWG, they had a sincere interest in data and data-driven policymaking, which likely influenced my expectations about how my research would be received. Rather than taking the time to discuss how scholars had historically been viewed in the business and human rights policymaking arena, clarify my role and the role of data, and better understand the UNWG's interest, I jumped right in.

Challenge 2: Sharing Unpopular Findings

A few years after the initial conversation with our UNWG contact, I had the opportunity to present our preliminary findings at two different briefings with the UNWG. One meeting was specifically about data gaps and how data could help better understand and improve business conduct regarding human rights. I was ecstatic about the opportunity to share some of our early findings, in part because some of them challenged underlying assumptions made by policymakers at the global level. The findings were intriguing, and I assumed others would think so too.

As I prepared my presentation, I was deep in the data and focused on what I thought were the most interesting results. As a young scholar, I paid less attention to policymakers' possible perspectives, their constraints, or which findings would be politically palatable. The research was political, as discussed below, but I did not see it as such (see also Erdberg Streadman, Chapter 9, this volume). This was, in part, because our conversations leading up to the meeting revolved around the dearth of systematically collected data on business and human rights and the UNWG's commitment to evidence-based policy; I assumed alignment. I did not fully appreciate the divergence between their stated ideals and the reality of the situation—that certain types of data and findings would simply not prompt further action. Now, I appreciate that there are different moments for different types of engagement, and, at least from the UNWG's point of view, this was *not* the moment to challenge their broader approach (even if they had been making calls to value and employ evidence-based policymaking). The UNWG was already on the path of implementing the UNGPs and did not want to hear information that contested the underlying assumptions of this approach.

The UNGPs assume that the "governance gap," or the combination of weak states and strong firms, is pervasive and persistent. The common narrative is that states are unable to hold companies accountable for human rights abuses because of state capacity issues or political will. The lack of political will was thought to be a driving force in the creation of the governance gap because increased trade and globalization have expanded the size, reach, and power of many multinational corporations. Thus, the UNGP's depiction of the "governance gap" suggests that victims of corporate human rights abuse have no access to formal remedy mechanisms at the domestic level. In 2008, Ruggie wrote:

> The root cause of the business and human rights predicament today lies in the governance gaps created by globalization—between the scope and the impact of economic forces and actors, and the capacity of societies to manage their adverse consequences. These governance gaps provide the permissive environment for wrongful acts by companies of all kinds without adequate sanctioning or reparation. How to narrow and ultimately bridge the gaps in relation to human rights is our fundamental challenge. (2008, para. 3)

Yet, the governance gap narrative overlooks the complexity of accessing justice for corporate human rights abuse (Deva 2011) and underestimates state actors in low- and medium-income countries (Bartley 2018). As I argue elsewhere (Olsen 2023), the governance gap has served as a useful heuristic at the global level, but it does not reflect what is happening empirically.

In contrast to showing that the governance gap was alive and well, the CHRD-LA illustrates that victims and their advocates *are* accessing remedy mechanisms more often than the governance gap might suggest (see Olsen 2023). The data illustrate that approximately one-third of victims were able to access some type of judicial or nonjudicial remedy mechanism, the latter of which can be led by state actors, community actors, or corporate actors. Certainly, impunity exists—victims are left without remedy in response to most allegations of corporate human rights abuse. Yet, I thought that data on hundreds of instances in which victims have access to some type of remedy process (even if the outcome of such processes falls short) would provide a strong foundation upon which the UNWG might shift their work toward improving access and increasing accountability. I thought the data might allow the UNWG to identify what practices were already being used, build upon successes, and learn from failures to promote more targeted policy rather than a blanket promotion of all types of remedy. The data on remedy outcomes, I imagined, would help them provide more nuanced recommendations and best practices to member states.

Second, a tacit assumption at the time (2011–2015) was that most democracies had moved beyond an era of state-sponsored abuse. Thus, the idea behind the business and human rights movement was that global human rights advocates should turn their focus to corporate actors so that they might begin reducing and redressing corporate human rights abuses. The human rights regime that developed post–World War II (e.g., the International Bill of Human Rights), combined with the "third-wave" of democratic transitions in the 1970s and 1980s (Huntington 1991), ushered in a new era in which states were aware of, and improving on, their human rights records, especially through transitional justice efforts (Teitel 2003; Olsen, et al. 2010). The assumption was that states would play an offensive role and help promote greater respect for human rights by corporations headquartered in, or operating within, their borders.

The CHRD-LA data, however, indicated that the state was either complicit, or had assisted, in approximately one-third of all allegations of abuse in the corporate context between 2000–2014 in Latin America. Elsewhere, another coauthor and I found that state involvement in corporate human rights abuse (what we term "economic complicity," as opposed to "corporate complicity") is more likely when a country's transition to democracy was negotiated with outgoing political elites, when economic rents depend on mineral extraction, and when union members are present (Olsen and Bernal-Bermúdez 2023). While there are numerous examples of economic complicity, a common occurrence is when community members or employees are protesting outside of a worksite and the police use excessive force to disperse the crowd. In many cases, individuals are forcibly detained, beaten, or murdered. In these circumstances, while the companies did not directly commit the abuses, they are implicated in the abuses or deaths of community members and are expected by external stakeholders to address the wrongdoing. We argue that legacies of violence persist in the democratic period and, in this case, permeate the corporate context in ways unexplored by business and human rights policymakers and scholarship.

When I presented this finding, I expected others to be surprised or astonished. At a minimum, I expected some type of engagement. Instead, the room fell silent. At the time, I knew, but did not fully appreciate that the very existence of the UNWG depended upon *states'* support. And, the UNWG's ability to ensure the UNGPs were adopted widely depended upon *corporate* cooperation. No one wanted to hear that states exacerbate abuses. Or that the state plays dual roles in both promoting greater respect for human rights from corporate partners while also abusing those rights on behalf of those same partners. In short, no one wanted to hear about the state's economic complicity (Olsen and Bernal-Bermúdez 2023).

Misaligned Incentives

For both sets of findings—first, that victims have access to remedy more often than many would expect, and, second, that the state engages in abuses or economic complicity in the corporate context—I failed to recognize or internalize the incentives of the UNWG. Even with our many conversations on data-driven policymaking, the findings I had to share did not align with their mandate. In psychological research, there is a term called "motivated inference," which describes the scenario in which people have strong motivations or incentives and are therefore very selective in the sort of evidence that they absorb or internalize (Kunda 1987). There are tendencies of self-preservation or self-interest at play that shape cognitive processing. As Kunda explains, "The motivation to be accurate enhances use of those beliefs and strategies that are considered most appropriate, whereas the motivation to arrive at particular conclusions enhances use of those that are considered most likely to yield the desired conclusion" (1990: 481). The UNWG members with whom I met (who have since finished their terms) had already embarked on a particular pathway of implementing the UNGPs to their fullest potential. There was no appetite on behalf of policymakers to engage with the complexity of the research or what it meant for their work (see also Epstein, Chapter 7, this volume). Nor was there much interest in reflecting upon the pathway they had already chosen.

What explains the drive to implement the UNGPs, and the reluctance to revisit its logic or appropriateness? The UNGPs are the result of a decades-long debate in the academic literature and in policy communities about the value and limits of soft- versus hard-law mechanisms to spur change—one that many, at the time, believed was a settled debate. As mentioned earlier, the breadth of the Draft Norms was met with trepidation. The main point of controversy was that they placed direct, legal responsibility on corporations as "organs of society" (a term that is employed in the Universal Declaration of Human Rights). The Draft Norms also included implementation mechanisms (monitoring, reporting, and verification) and stipulated obligations of financial redress (Weissbrodt and Kruger 2003). Businesses largely opposed the Draft Norms as they wanted to avoid binding international law.

States fought this hard-law initiative, as well. Many developing countries questioned their ability (and willingness) to enforce a regulatory framework that was perceived as threatening by the businesses upon which their economies depended. Developed states were hesitant to sign on to hard-law mechanisms that would challenge state sovereignty through extraterritoriality. Even so, in August 2003, the UN sub-Committee on the Promotion and Protection of Human Rights approved the Draft Norms. Yet, as the political backlash grew, it led to the UN Commission on Human Rights setting the Draft Norms aside in 2004, declaring they "had not been requested by the commission and thus, would not be considered" (Mantilla 2009).

Instead of pushing hard law, Ruggie sought voluntary compliance through the promotion of global norms (Ruggie 2013). He completed nearly three years of meetings with various advocacy organizations, member states, and company leaders to garner support for the initiative. By creating a focal point and convening public, private, and civil society actors, the UNGPs could augment existing efforts to improve compliance and respect for human rights in the corporate context. It also created broad norms that facilitate bringing corporate actors into the fold, which some scholars suggest constitutes a norms cascade with the potential to impact business behavior (Deitelhoff and Wolf 2013; Mwangi, Reith, and Schmitz 2013). Ruggie tacitly drew from Ostrom's work (Ostrom 2010) and suggested that the creation of the UNGPs was an exercise in polycentric governance. This approach also reflected Ruggie's scholarly work on new governance theory, which "rests on the premise that the state by itself cannot do all the heavy lifting required to meet most pressing societal challenges" (Ruggie 2014: 8).

In sum, there was little incentive for members of the UNWG to overcome their motivated inference given the extraordinary work involved in creating momentum, agreement, and "thick stakeholder consensus"

(Pauwelyn, Wessel, and Wouters 2012: 1–2) around the UNGPs. The UNWG was wholeheartedly committed to ensuring the UNGPs were a "success." Their interest in implementing the UNGPs was strong, not only to improve corporate conduct for human rights but to also to ensure the legitimacy of the UNGPs themselves and to avoid hard-law efforts that were considered a dead end. Our incentives were simply misaligned.

Challenge 3: Fostering Conversation amid Contention

I did not have the opportunity, nor did I reach out, to share additional findings with the UNWG in additional closed-door meetings.

Misaligned incentives can partly explain this, but the UNWG also became quickly consumed with the debate over whether there should be a treaty on business and human rights. Only three years after the UN Human Rights Council endorsed the UNGPs, a movement was afoot to deepen the binding and enforceable global norms around business and human rights. The Council narrowly passed a resolution to create an Open-ended Intergovernmental Working Group (IGWG) to draft a treaty on business and human rights. Promoted primarily by Ecuador, South Africa, Bolivia, Cuba, and Venezuela, the so-called "Zero Draft" has gained momentum (Ford and O'Brein 2017). The UN Human Rights Council heard the first read in October 2018, while discussions continue and the tenth session of the IGWG was held in October 2024. Some of the strongest proponents of the treaty efforts perceive it as a necessary alternative path, as the UNGPs lack legally binding obligations to hold corporations (or the states in which they are headquartered) accountable for wrongdoing (Kamminga and Zia-Zarifi 2000; Weissbrodt and Kruger 2005; Deva 2017). The UNGP is now comprised of members who are engaging directly with the treaty debate while simultaneously trying to ensure that, should the treaty move forward, it supports rather than supplants the UNGPs.

Even though the move away from data was driven by other changes within the UN, this dilemma still raises the following question: when researcher–practitioner incentives are misaligned, how does one challenge an institution and its stakeholders when their good will is necessary for bringing about better policy outcomes?

There are several ways I could have shared these findings and my work differently. First, it is important to frame unpopular findings at the beginning of a conversation so that there is space to discuss why the findings might be unpopular or what makes them challenging to current efforts. I am reminded of a relatively nascent literature in political theory on agonism (see Honig 1993; Mouffe 2013) that challenges the notion inherent in deliberative democratic theory that consensus is desirable, as lack of consensus may threaten the self-preservation of the political unit in question (e.g., the party, a committee, or working group). Agonism, alternatively, suggests that consensus is always suboptimal as all decisions exclude some portion of the population. Decisions, in turn, may create new divisions and/or motivate previously unorganized factions to come forth to contest the decision that was just made. Rather than lamenting contestation, an agonistic approach suggests that legitimacy can be gained through contestation (see Olsen 2023; Olsen and Van Buren 2023). To achieve that, however, unpopular findings should be presented as a potential starting point for a deeper conversation in which contestation is not perceived as threatening, but as an opportunity for greater learning.

On a more practical level, this might also be a case of missing the "meeting before the meeting." While I had the opportunity to engage with my contact at the UNWG, I had not reached out to other members of the working group and was unsure if it would have been appropriate to do so. Had there been an opportunity to socialize the findings with other UNWG members and learn more about the constraints they faced prior to my presentation, I imagine I would have been able to engage more deeply with the UNWG. This is a challenge, addressed elsewhere in this volume, of being the "external" expert (see Erdberg Steadman, Chapter 9, this volume).

Another anecdote reiterates my unique position as an academic relative to the constraints confronting the UNWG and NGO advocates. In one of the meetings, the group began brainstorming how the UNWG might use data to engage with other stakeholders. There was some interest in using data to create indices that could track performance of states and companies to provide some type of assessment of their business and human rights record or implementation of the UNGPs. Representatives from NGOs initially expressed skepticism. They did not want to "name names," as they were actively creating partnerships and working with some of the world's largest companies. Having brought these companies into the fold, they did not want to risk that relationship by creating an index that could deter their further engagement by calling these corporations out for human rights abuses.

Those within the UNWG also communicated their reservations. They did not want to shame states that were engaging in the agenda, but perhaps doing less than hoped. The UNWG was beholden to states for a variety of reasons. They wanted to—indeed, needed to—engage with states and facilitate their promotion of the UNGPs to improve overall respect for human rights in the corporate context. At this point, I stated that as an academic I am not beholden to states or companies. I would be able to share both best *and* worst practices of states *and* companies, including the ones that were behaving poorly or not meeting expectations. One member of the UNWG chuckled and said, "Well, we will refer to you as our member with impunity." This comment, I believe, is indicative of the confines of UNWG's work and, perhaps, their inability to see how they could engage with data or findings that challenged the UNGPs. Had I been more aware of these internal dynamics, I could have framed the findings in a way that empowered them to act, rather than push away complexity.

To be clear, this account is not meant to denigrate or criticize the work of the UNWG or any others advocating for improved corporate behavior or victims' access to remedy. It simply highlights that resources are often spent defending one camp or another, rather than understanding how to better coordinate and collaborate on existing efforts. John Dewey once wrote that today's institutions are the "residue" of yesterday's problems (1927: 74). If we cannot embrace challenges to the status quo and innovate as a result, especially in complex policy areas like business and human rights, we might find ourselves simply adding to the residue.

Assessing the Engagement: Misaligned Incentives and the Policy Cycle

Reflecting on the challenges of ethical engagement with policymakers, I am reminded that engagement plays different roles at different moments in the policymaking cycle. Perhaps simply getting my findings on the UNWG's radar was all that was possible at the time. My expectations around my engagement may have been too high, given the political context in which the UNWG found itself (e.g., with treaty debates) while also trying to encourage states and companies to embrace the UNGPs. I had hopes of shaping policymaking, whereas, given our misaligned incentives and timing, the UNWG was unable to reassess or consider the assumptions inherent in its work (also see Erdberg Steadman, Chapter 9, this volume).

A second reflection, inspired by the literature on agonism, makes me realize that the singular focus on ensuring widespread adoption of the UNGPs (e.g., seeking to solidify some sort of consensus) also prohibited members from exploring how my findings could work in tandem with, or strengthen, the UNGPs. They were unable to embrace the contestation associated with findings that challenged certain assumptions made in the creation of the UNGPs and learn from them to envision a myriad of possible approaches to improve human rights.

A final reflection is to understand that when academics have research agendas that are relevant to policymakers, it is important to keep the long game in mind. While I was hopeful my engagement with the UNWG on the powerful role data can play to inform the business and human rights agenda would have been deeper, I did the best I could based on what I knew at the time. Engagement with policymakers in a specific policy space, however, is not a one-shot game. Making meaningful contributions means additional doors will open. I was subsequently invited to present on a panel at the UN Forum on business and human rights, which draws thousands of advocates, state representatives, and academics annually. In 2020, I coauthored a white paper using our database on victim access to remedy for corporate human rights abuses in Latin America, which I specifically prepared for the Office of the United Nations High Commissioner for Human Rights (OHCHR), and which is now posted on their website (see Olsen, et al. 2020).[4] Additionally, I continue to keep in contact with members of the UNWG and individuals working on access to remedy, specifically, in the OHCHR. I am grateful for these additional opportunities and ongoing conversations.

Conclusion

This case highlights some of the challenges of engaging with policymakers when incentives are not aligned. The case raises questions about how scholars can clarify their role, share unpopular findings, and foster conversation amid contention, and hopefully sparks some thoughts from others who have found themselves in uncomfortable positions when engaging with policymakers.

As scholars enter into policy engagements, they might consider several key questions. First, the relationship between policymakers and academics evolves over time. How can clear expectations be communicated on both sides when the research and the political dynamics of policymaking are changing simultaneously? Second, the lack of an opportunity to socialize findings can be a constraint. What can academics do to facilitate greater engagement when policymakers determine both the way in which they

[4] See: https://www.ohchr.org/sites/default/files/Documents/Issues/Business/ARP/Olsen_Bridging_Data_Gap_January2020.pdf.

participate and with whom? Third, how does one criticize an institution and its stakeholders when their good will is necessary for bringing about better policy outcomes? What obligations (if any) do scholars have when policymakers do not care to hear the message?

My hope is that reflections on this case will help other scholars anticipate the complexities and social pressures they might encounter if they find themselves presenting unpopular findings. Carefully considering the difference between one's interlocutors' expressed interests and what they are politically capable of doing is critical. While there is certainly no strict guide for engagement as the case studies throughout this volume attest, being aware of possible misaligned interests in advance can assist researchers in navigating the policy process. My aim has been to help scholars create space to discuss challenging findings in a way that will further engagement, rather than stymie it.

References

Bartley, Timothy (2018) *Rules without Rights: Land, Labor, and Private Authority in the Global Economy*. Oxford: Oxford University Press.

Dancy, Geoff, Bridget E. Marchesi, Tricia D. Olsen, Leigh A. Payne, Andrew G. Reiter, and Kathryn Sikkink (2019) "Behind Bars and Bargains: New Findings on Transitional Justice in Emerging Democracies." *International Studies Quarterly* 63(1): 99–110.

Deitelhoff, Nicole, and Wolf, Klaus Dieter (2013) "Business and Human Rights: How Corporate Norms Violations Become Norm Entrepreneurs." In *The Persistent Power of Human Rights*, ed. Thomas Risse, Stephen C. Ropp, and Kathryn Sikkink. Cambridge: Cambridge University Press, pp. 222–238.

Deva, Surya (2011) "Protect, Respect and Remedy: A Critique of the SRSG's Framework for Business and Human Rights." In *Corporate Social and Human Rights Responsibilities*, ed.Karin Buhmann, Lynn Roseberry, and Mette Morsing. Houndmills/Basingstoke: Palgrave Macmillan, pp. 108–126.

Deva, Surya (2017) "Scope of the Legally Binding Instrument to Address Human Rights Violations Related to Business Activities," https://www.escr-net.org/sites/default/files/scope_of_treaty.pdf

Dewey, John (1927) *The Public and Its Problems*. Athens, OH: Swallow Press.

Ford, Jolyon, and Claire Methven O'Brien (2017) "Empty rituals or workable models? Towards a business and human rights treaty." *The University of New South Wales Law Journal* 40(3): 1223–248.

Honig, Bonnie (1993) *Political Theory and the Displacement of Politics*. Ithaca, NY: Cornell University Press.

Humphrey, Michael (2003) "International Intervention, Justice and National Reconciliation: The Role of the ICTY and ICTR in Bosnia and Rwanda." *Journal of Human Rights* 2(4): 495–505.

Huntington, Samuel P. (1991) "Democracy's Third Wave." *Journal of Democracy* 2(2): 12–34.

Kamminga, Menno T. and Saman Zia-Zarifi (2000) "Liability of Multinational Corporations under International Law: An Introduction." In *Liability of Multinational Corporations under International Law*. The Hague: Kluwer Law International, pp. 1–15.

Kunda, Ziva (1987) "Motivated Inference: Self-Serving Generation and Evaluation of Causal Theories." *Journal of Personality and Social Psychology* 53(4): 636.

Kunda, Ziva (1990) "The Case for Motivated Reasoning." *Psychological Bulletin* 108(3): 480.

Mantilla, Giovanni (2009) "Emerging International Human Rights Norms for Transnational Corporations." *Global Governance: A Review of Multilateralism and International Organizations* 15(2): 279–298.

Méndez, Juan E. (1997) "Accountability for Past Abuses." *Human Rights Quarterly* 19(2): 255–282.

Mouffe, Chantal (2013). *Agonistics: Thinking the World Politically*. London: Verso Books.

Mwangi, Wagaki, Lothar Rieth, and Hans Peter Schmitz (2013) "Encouraging Greater Compliance: Local Networks and the United Nations Global Compact." In *The Persistent Power of Human Rights*, ed. Thomas Risse, Stephen C. Ropp, and Kathryn Sikkink. Cambridge: Cambridge University Press, pp. 273–301.

Olsen, Tricia D. (2023) *Seeking Justice: Access to Remedy for Corporate Human Rights Abuse*. Cambridge: Cambridge University Press.

Olsen, Tricia D., Leigh Payne, and Andrew Reiter (2010) *Transitional Justice in Balance: Comparing Processes, Weighing Efficacy*. Washington, D.C.: USIP Press.

Olsen, Tricia D. and Laura Bernal-Bermúdez (2023) "Uncovering Economic Complicity: Explaining State-led Human Rights Abuses in the Corporate Context." *Journal of Business Ethics* 189(1): 35–54.

Olsen, Tricia D. and Harry J. Van Buren (2023) "Agonistic Respect and the Ethics of Employment Relationships." *Journal of Business Ethics*, 1–14.

Olsen, Tricia D., Brittny Parsells-Johnson, Laura Bernal-Bermúdez, Michelle Westermann-Behaylo, and Leigh A. Payne (2020) "Bridging the Data Gap: Exploring Pillar III and Victims' Access to Remedy." White Paper prepared for the Office of the United Nations High Commissioner for Human Rights. https://www.ohchr.org/sites/default/files/Documents/Issues/Business/ARP/Olsen_Bridging_Data_Gap_January2020.pdf.

Olsen, Tricia D., Kathleen Rehbein, Annie Snelson-Powell, and Michelle Westermann-Behaylo (2022) "Human rights in the oil and gas industry: When are policies and practices enough to prevent abuse?" *Business & Society* 61(6): 1512–1557.

Osiel, Mark J. (2000) "Why Prosecute? Critics of Punishment for Mass Atrocity." *Human Rights Quarterly* 22: 118.

Ostrom, Elinor (2010) "Beyond Markets and States: Polycentric Governance of Complex Economic Systems." *American Economic Review* 100(3): 641–672.

Pauwelyn, Joost, Ramses Wessel, and Jan Wouters. eds. (2012) *Informal International Lawmaking*. Oxford: Oxford University Press.

Payne, Leigh. A., Gabriel Pereira, and Laura Bernal-Bermúdez (2020). *Transitional Justice and Corporate Accountability from Below: Deploying Archimedes' Lever*. Cambridge: Cambridge University Press.

Ruggie, John G. (2008) "Protect, Respect and Remedy: A Framework for Business and Human Rights." Human Rights Council's Report of the Special Representative of the Secretary-General on the Issue of Human Rights and Transnational Corporations and Other Business Enterprises. April 7. UN Doc. A/HRC/8/5, para. 3. https://media.business-humanrights.org/media/documents/files/reports-and-materials/Ruggie-report-7-Apr-2008.pdf.

Ruggie, John G. (2013) *Just Business: Multinational Corporations and Human Rights*. New York: WW Norton & Company.

Ruggie, John G. (2014) Global Governance and New Governance Theory: Lessons from Business and Hhuman Rights. *Global Governance* 20: 5.

Teitel, Ruti G. (2003) Transitional Justice Genealogy. *Harvard Human Rights Journal* 16: 69.

Weissbrodt, David and Muria Kruger (2003) Norms on the Responsibilities of Transnational Corporations and Other Business Enterprises with Regard to Human Rights. *American Journal of International Law* 97(4): 901–922.

Weissbrodt, David and Muria Kruger (2005) "Human Rights Responsibilities of Businesses as Non-State Actors." In *Non-State Actors and Human Rights*, ed. Phillip Alston. Oxford: Oxford University Press, pp. 315–350.

4

Coping with Conflicting Claims

Ethnic Power Sharing for Sustainable Peace?

Timothy D. Sisk

In late 2007 and into early 2008, Kenya was gripped in escalating violence as the country underwent presidential elections which split the electorate principally along ethnic lines. Violence escalated before the election and afterwards. More than 1,300 lives were lost, and some 600,000 persons were displaced in the increasingly ethnic confrontations in the streets and in regional metropoles such as Kisumu in the Rift Valley. To mediate, a Panel of Eminent African Personalities, chaired by former United Nations Secretary-General Kofi Annan, stepped in. Over forty-one days of talks, the mediators induced the Kenyan protagonists to adopt a broad-based political power-sharing settlement that averted the dangerous and devastating slide into "ethnic" civil war.[1]

In pursuing mediation, Annan and the African eminent persons' group faced a dilemma. They could *encourage* the parties to adopt an ethnic power-sharing arrangement based on the social mobilization realities on the ground and recognize ethnic representation, territorial decentralization, and minority veto. Alternatively, they could *avoid* an elite-based bargain that cements ethnicity in politics and instead seek to encourage the crosscutting ties and social cohesion needed to sustain peace (United Nations Development Program 2020).[2]

Annan had seen such "ethnic" power sharing," or institutions that guarantee ethnic groups specific rights and proportional ethnic representation, as global mediator in Lebanon (1989), Bosnia (1995), Northern Ireland (1998), Zimbabwe (1998), and Iraq (after 2003). Yet such ethnic pacts are the subject of strong practitioner and scholarly debates, with many competing

[1] For details on the Kenya mediation described here, see Lindenmeyer and Kaye (2009) and Cohen (2008).

[2] The dilemmas of power sharing that engender social traps that perpetuate ethnic politics are described well in Jarstad (2008).

Timothy D. Sisk, *Coping with Conflicting Claims*. In: *Speaking Science to Power*. Edited by: Rachel A. Epstein and Oliver Kaplan, Oxford University Press. © Timothy D. Sisk (2024). DOI: 10.1093/oso/9780198875185.003.0004

claims. There are claims that, in efforts to "stop the killing," policymakers should encourage the adoption of ethnic power-sharing arrangements to assure the parties. US diplomat Richard Holbrooke used coercive diplomacy to push for a power-sharing formula in Bosnia-Herzegovina in 1995 to prevent further bloodshed (Holbrooke 1998). Others decry such ethnic power-sharing pacts and claim they frustrate cohesion and a shared or common destiny—the underlying social basis of peace—since they may institutionalize ethnic elites.[3] In the longer term, power sharing in Bosnia seems to have entrenched ethnicity, setting the stage for new conflict escalation (Andrés 2021). In Kenya, Annan and colleagues had learned from such examples and opted to steer Kenya away from a pact rooted in ethnicity.

The cases of postconflict political transitions raise the following question: Can ethnic power sharing and inclusive political institutions facilitate sustainable peace and, if so, under what conditions? Both policymakers and scholars have disagreed on the answer, generating competing claims. This chapter explores the responsible engagement challenges that arise when there are competing claims in policy-relevant areas of social science research. Such contention over scientific results is a natural part of research and discovery, shaping the gaps and puzzles that spur further research. Yet they can also create a confusing environment and steer policymakers astray. I argue that, in such circumstances, there are several risks for scholars as they seek to communicate research insights to policymakers, with the resulting possibility of arriving at suboptimal policy decisions.

This chapter focuses specifically on the challenges posed by some of the competing claims surrounding the efficacy of ethnic power sharing to sustain peace. These include the controversies over so-called "consociational" versus "centripetal" approaches to designing political institutions, as well as debates over several power-sharing sub-elements such as electoral system reform and decentralization. The former approach recognizes ethnicity and seeks to structure inclusion along these lines, whereas the latter emphasizes the importance of institutions that eschew ethnic representation as the basis of the state.

[3] This formulation of the risks of ethnic power sharing in sustaining peace are drawn from the 2008–2009 UNDP National Human Development Report for Lebanon, titled "Toward a Citizen State" which cataloged the development and social-conflict consequences of the ossified ethnic power sharing arrangements adopted in the internationally mediated Ta'if Accord of 1989 (National Reconciliation Agreement) and reaffirmed in the Qatar Agreement of 2008.

As researchers and policymakers sort through competing claims, there are several steps they can take toward responsible behavior, even when answers are not evident and they face epistemic uncertainty. I enumerate several options. Scholars can conduct additional research to resolve competing claims. They can also broaden their methodological tool kits to identify for policymakers which generalizations are highly contingent on conditions and to articulate under what conditions "power sharing" features can contribute to immediate- or longer-term efforts to sustain peace. They can provide information about potentially influential "unknown" factors and highlight the short- and long-term trade-offs and consequences of policy choices. They can use "strategic imagination" to identify and articulate best- and worst-case scenarios. Finally, they can learn about and increase awareness of policymakers' frames for processing conflicting claims.

The chapter begins by outlining the urgent need for—and usefulness of—power-sharing solutions to help ameliorate contemporary identity-based conflicts. Second, it identifies the key competing claims among scholarly literature and policymaker constructs on the topic of ethnic power sharing. Third, it discusses some possible guidance for how scholars and policymakers can cope with competing claims to arrive at better-informed decisions. The conclusion broadens the insights about competing claims and summarizes productive modes for future action and analysis. It argues that responsibly engaged scholars should move from advocating generalized models to being honest brokers who explore the contingent conditions under which such models may lead to sustainable peace.

Agreement on the Urgency and Utility of Inclusive Power Sharing

While there is disagreement among scholars on the institutions that can help sustain peace through power sharing, there is largely consensus on the utility of inclusivity in governance in some form. More specifically, there is consensus on the dangers of majoritarianism, or rule by a majority coalition. These include the potential distortions in vote-to-seat outcomes, the inability of geographically dispersed minority parties to achieve representation, and—in the context of an ethnic party system—the likelihood that a single ethnic group or coalition of ethnic groups will govern exclusively and to the detriment of others.

Majoritarian systems seem to be especially vulnerable to elites who "play the communal card" as an instrumental strategy, sowing fear and trumpeting nationalism as a deliberate electoral strategy.[4] Arend Lijphart, the most indefatigable critic of majoritarian and plurality electoral rules for divided societies, refers to the potential for "majority dictatorship" (Lijphart 1977: 25–28, 114–118). Donald Horowitz concurred in his seminal work *Ethnic Groups in Conflict*, arguing that under conditions of simple majority rule

> Ethnic parties developed, majorities took power, and minorities took shelter. It was a fearful situation, in which the prospect of minority exclusion from government underpinned by ethnic voting, was potentially permanent . . . Civil violence, military coups, and the advent of single party regimes can all be traced to this problem of inclusion-exclusion. (Horowitz 1985: 629)

Scholars generally agree that majority-rule democracy is harmful to peace in situations of deep social diversity, such as Sri Lanka.[5] The consensus in recent research contributes to a belief in the prospects for political or constitutional engineering to mitigate conflicts. There are solutions such as designing incentives for inclusivity in political parties and for focusing on accountability and responsive government over the advancement of identity (Reilly and Norlund 2008).[6] Many policymakers therefore routinely reject majoritarian rule institutions in societies with strong patterns of ethnic political mobilization, as electoral processes may become simply "ethnic censuses" with little hope of minority inclusion over time.[7]

At the same time, the threats to peace—and the need for tested policy solutions—are as urgent now as ever. Identity-related conflict worsened in the late 2010s and, in 2021, the Minority Rights Group (MRG), in its annual *Peoples under Threat* global survey, reported persistent identity-related conflicts and multiple threats to minorities and indigenous peoples worldwide (Minority Rights Group 2021). This re-opens longstanding scholarly and policy-oriented questions about designing inclusive democracy

[4] For an earlier analysis of the ethnic entrepreneur problem as a driver of violence in the early 1990s, see Human Rights Watch (1995).

[5] In the case of Sri Lanka, there is broad agreement that simple majoritarianism works against inclusivity and may even lead to domination by majorities and seemingly permanent exclusion of minorities (AP 2020; DeVotta 2021). However, in recent years some countries have emerged from conflict and done well in contexts of majoritarian democracy, such as Liberia's three rounds of majoritarian presidential elections since the Second Liberian Civil War ended in 2005.

[6] Examples of such regulation include party membership and registration requirements, electoral system choice, and rules on "party hopping" or switching parties in between elections.

[7] For an analysis of these dynamics in the run-up to the May 2018 elections in Iraq, see Coker and Hassan (2018).

for interethnic harmony in societies riven by identity-based politics. What approach to power sharing is best for which situations?

But What Approach? The Competing Claims about Ethnic Power Sharing

Some scholars see power sharing along ethnic lines as desirable or inevitable given the hardening of identity over time. In 1977, Arend Lijphart's landmark book *Democracy in Plural Societies* articulated a model of sharing power along ethnic lines (known as "consociationalism"). It envisioned grand coalitions, proportional distribution of power, and ethnic political parties that are ostensibly consistent with principles of democracy—principles previously associated mostly with "majority rule" (Lijphart 1977). Other scholars point to the social-trap risks of power sharing and argue against these solutions.

Beneath this debate are several competing claims that are partially rooted in differences in research methodologies. Large-N studies tend to show that power-sharing institutions and practices are conflict-mitigating over time. By contrast, some case-specific experiences highlight the pathologies of power sharing along identity lines, as in contexts such as Bosnia, Northern Ireland, Iraq, and Lebanon. Yet in still other cases, such as South Tyrol, ethnic power sharing is seen as a successful mechanism of conflict regulation (Wolff 2004).

"Power sharing" is not a single formula. Protagonists and outside mediators must make difficult choices over several alternative institutional configurations to craft an inclusive political settlement.[8] Power sharing is thus more of a spectrum of options rather than a single formula or set of specific political institutions like proportional representation or federalism (Harris and Reilly 1998, Hartzell and Hodie 2007, Roeder and Rothchild 2005). There are also disagreements among scholars over the effectiveness of specific power-sharing institutional designs for conflict-mitigating outcomes. I delineate the most important areas of the debate and review the competing advocacy claims of the scholars—and the consequences—surrounding them. These include the debate between "consociationalists" and "centripetalists," debates over electoral system choices, the recurring contestation over federalism,

[8] Recent research projects on power sharing include the special issue of UNESCO's *International Journal on Multicultural Societies*, "Democracy and Power-Sharing in Multi-National States" (Volume 8, Number 2 (2006), and the extensive work of the German Institute for Global and Area Studies (GIGA) project "Institutions for Sustainable Peace," available at https://www.giga-hamburg.de/en/research-and-transfer/projects/institutions-sustainable-peace-comparing-institutional-options-divided-societies-post-conflict-countries.

and the extent to which ethnicity or ethnic parties and politics should be recognized and more fixed or fluid, and even disincentivized.

Debating "Consociational" Power Sharing and Centripetalism

One side of the debate on power sharing emphasizes guaranteed minority inclusion and, in essence, a minority "veto" over the most critical issues for a society. To mitigate the dangers of permanent minority exclusion under majority rule, the term "power-sharing" has been promoted as a solution. It has been defined by Lijphart as a set of democratic principles that—when carried out through practices and institutions—provide each significant identity group or segment in a society with representation and decision-making abilities on common issues and a degree of autonomy over issues of importance to the group. Lijphart's principles of power-sharing—known as "consociational democracy" (from the Latin term *consociatio*, to associate in an alliance) was pathbreaking in its differentiation of coalescent democracy from majoritarian democracy.[9]

Consociationalism, above all, relies on elite cooperation for successful conflict management (Lijphart 2004). Examples include war-ending peace agreements such as the 1989 Ta'if Accord in Lebanon, the 1995 Dayton Accords in Bosnia, and the 1998 Belfast Agreement in Northern Ireland. Consociationalists suggest that even if there are deep communal differences, overarching integrative elite cooperation is a necessary and sufficient condition to assuage conflict. In the consociational approach, elites, or conflict group leaders, directly represent various societal segments and act to forge political ties at the center.

The debate over the effectiveness of consociational approaches is complicated, however, as both in theory and in practice they have been criticized as nondemocratic (forming a government before electing it), reinforcing of ethnicity and nationalism at the expense of national unity, disadvantageous for women's representation, prone to gridlock, and rewarding of ethnic patronage networks. As Donald Horowitz (2014) insightfully describes, ethnic power sharing is beset by three fundamental problems: when can such systems be adopted by parties in good faith as a way to live together, how

[9] Consociationalism relies on four elements or principles (as the precise institutional form may vary): inclusive executives along the lines of a "grand coalition"; proportionality in elections, public appointments, and fiscal matters; territorial autonomy or nonterritorial (group rights) autonomy, sometimes known as "corporate" autonomy; and mutual or minority veto such that minorities may exercise self-determination over matters of importance to the group, such as language and education policy.

can democracy survive when parties are based on ethnic politics, and what happens when ethnic elites at the center fail to agree or moderate?

In each of the cases of Bosnia, Lebanon, and Northern Ireland, consociationalism has led to a short-term, "cold" peace but has engendered the long-term perpetuation of identity-based politics and ethnic political parties (see Aoun and Zahar 2017 and Calfat 2018 on Lebanon). It is thus unclear under what conditions the rigid ethnic-based aspects of consociational power-sharing can be adopted through incremental reform (the so-called sequencing dilemma, relevant to the introduction apropos of Kenya). Critics of consociationalism question whether its representation of groups *as groups* leads to better conflict management than alternative approaches.

Centripetalism has arisen as the main alternative approach to consociationalism. This approach seeks to foster political organizations that are integrative, or which transcend ethnic group differences (Wimmer et al. 2004; McCulloch and McGarry 2017). It espouses the creation of pre-election coalitions between ethnic parties or the creation of broad multiethnic parties around interests such as region or economics (a centripetal/inward-spinning approach) (Horowitz 1985: 597–600). It also involves inducements for interethnic cooperation, such as electoral laws that effectively promote pre-election electoral coalitions through vote-pooling.[10] Centripetalism can be seen in contexts such as Nigeria, where electoral rules for the presidency require broad-based political coalitions, where the proliferation of states over time has broken down ethnic territorialization, and where ethnic political parties are disallowed (McCulloch 2012). At the same time, centripetalist incentives-based mechanisms have been criticized for the lack of clearly successful examples that are effective in reducing ethnic tensions. In many instances, countries have adopted a mix of mechanisms that further highlights the gap between the general-theory seeking scholarship and the nitty-gritty details of political institutions in various highly specific contexts.[11]

[10] Horowitz (1985): 597–600. Some mostly consociational contexts such as Northern Ireland sought to soften the rigidity of ethnic representation through the adoption of the single-transferable vote, for example. A third component recommends policies to encourage alternative social alignments, such as social class or territory, by placing political emphasis on crosscutting cleavages; for example, through ministries devoted to the advancement of social cohesion, as mentioned in the Kenya example in the Introduction to the chapter. This approach features measures to reducing disparities between groups through managed distribution of resources. See Stewart (2000). Increasingly, scholarship in this area has emphasized the reduction of horizontal inequalities and the amelioration of multidimensional poverty, and a focus on peripheral or highly affected subregions.

[11] See, for example, Bogaards (2019). Please see the entirety of this volume of the SPSR for further analysis of consociationalism in several case studies.

In application, each type of institution may present compelling choices for policymakers, and the literature further provides competing claims and evidence. Has open-list PR (proportional representation) in South Africa contributed to the demise of a moderate center and an ethos of nonracialism and inclusion? Can territorial autonomy work to bring a definitive end to conflict in the Philippines? How effective will reserve seats be for the insurgent Revolutionary Armed Forces of Colombia (FARC) in the Colombian peace process? Can Iraq move beyond sectarianism in its ruling coalition? Can Nepal succeed as its society transformed from a history of "ranked" group interactions to one of inclusion and remediating historical disadvantages, all in the context of new ethnic mobilizations? What options exist to resolve the language crisis in Cameroon based on autonomy solutions seen elsewhere? How will Western societies grapple with increased multiculturalism and the rise of exclusive populism, focusing on formal (citizenship and naturalization processes) and informal (consultative or interim) means of inclusion?[12] Policymakers might ask whether in power-sharing contexts such as Bosnia or Lebanon it might be possible to move beyond consociationalism and toward more integrative solutions.

I assess four aspects of these competing claims on "best" approaches to sharing power to facilitate peace. In each of these, responsible engagement means recognizing that there is a high degree of conditionality in the application of these mechanisms. Scholars should be sure to articulate the conditions under which their claims are expected to hold up in complex contexts.

Debate 1: "Upstream Inclusivity" in Peace Processes: Exclusion from the Start?

A first contested subelement of the power sharing debate concerns the degree of inclusion in politics and peacemaking.[13] How should inclusive peace processes be practically structured, and how can such "upstream" inclusivity be best promoted? Research has shown that greater involvement of civil society in peace negotiations "upstream" leads, in a path-dependent way, to more durable peace over time (Paffenholz 2009; Nilsson 2012). For instance, the linkage between women's participation in peace processes and more peaceful politics is frequently advanced, with the causal mechanisms

[12] See, for example, the work of the Raoul Wallenberg Institute on inclusive societies at http://rwi.lu.se/what-we-do/focus-areas/inclusive-societies/, and International IDEA (2017).

[13] See the work of the project on political settlements at http://www.politicalsettlements.org/.

found in women's ability to build "bridging" social cohesion.[14] Landmark global norms, such as UN Security Council Resolution 1325 and the overall Women, Peace, and Security (WPS) agenda, have highlighted inclusivity along gender lines as an essential element in remaking relations across identity groups (Hellmüller and Zahar 2018).

However, some scholars argue that a political settlement among key elites is needed to guarantee peace and stability, with broadening of inclusivity (or not) over time (Jarstad 2008). The argument for keeping coalitions narrow in peacemaking may be aimed at reducing "spoilers" who may disrupt the process, as even in celebrated peace processes such as South Africa or Nepal elite dynamics have mattered most. This focus on elite dynamics echoes the earlier findings of the "transitology" literature in democratization studies that elite bargains are an essential stepping-stone toward greater inclusivity (Ould-Mohamedou and Sisk 2017). Yet they are also limited in that they tend to paper over local-level experiences and insecurities or provide for local-level security provisions (Mehler 2009).

The implications of this debate for responsible engagement are clear. While upstream inclusion in peace processes may lead to more inclusive political institutions, at the end of the day, political settlements must address the underlying short-term security considerations of the protagonists, in addition to long-term thinking that, for instance, places priority on women's representation.

Debate 2: Rethinking Presidentialism?

Presidentialism as a solution to manage social cleavages is also contested. The "perils of presidentialism" (as raised decades ago by Juan Linz [1990]) is a recurring debate in the literature on "neo-patrimonialism," and presidentialism-related crises capture news headlines around the world.[15] In the Democratic Republic of the Congo, Burundi, Gabon, Rwanda, and Uganda in Africa, for example, efforts to eliminate term limits and allow "presidents for life" have been variously successful or stimulated devastating conflict. Presidencies in Zimbabwe and South Africa have ended with

[14] As Tamaru and O'Reilly (2018: 1) describe: "Women repeatedly bridged divides in the negotiating process, contributing to peacebuilding and reconciliation in deeply divided societies, while also advancing consensus on key issues. They broadened societal participation and informed policymakers of citizens' diverse priorities for the constitution, helping to ensure greater traction for the emerging social contract."

[15] Observers point to the "degradation" of the presidency in the United States, for example; see Reich (2017). Brazil's crisis of democracy in 2022–2023 is also a problem of presidential executive aggrandizement, and presidentialism is arguably a strong driver of democratic backsliding in the Americas; see Pérez-Liñán, Schmidt, and Vairo (2019).

changes of regime in the context of misrule or corruption. Presidentialism can be fraught with problems: vulnerability to corruption, a tendency toward electoral manipulation and fraud, cooptation or capture of the military, and potential political party-weakening through patronage networks. By centralizing executive authority in one individual, such systems are more exclusive than parliamentary systems.

The conditions under which presidential systems may have conflict-exacerbating or conflict-ameliorating effects are highly complex and depend on factors such as geography and demographics, the quality of leadership, and the design of the electoral system. This produces a tendency toward overgeneralization and lack of case-specific analysis. Policymakers may be hamstrung by these debates, caught between the competing claims of academics and the power-hungry preferences of local actors, and reaching for presidential systems when they perceive they have a chance of winning a majority to avoid sharing power through more inclusive institutions. Rather than pushing alternatives for which there is unclear evidence, policymakers frequently default to local elite preferences.

Debate 3: Electoral System Choice: How Should Votes Become Seats?

Third, electoral systems are debated as a solution since they may variably encourage political inclusion or be subject to manipulations that restrict democracy. The electoral system is the formula by which votes are aggregated in a democracy to determine the winners and losers of seats in an assembly or office-holding. Scholarship on electoral systems focuses on the variety of systems available, their effects, and the mathematical conversions that occur to integrate votes across various systems to determine the winners and losers. There are competing claims over majority-rule systems and candidacy-based mechanisms versus proportional representation, and whether mixed-member or parallel systems and options such as rank-choice voting can have dampening or exacerbating effects on conflict.[16]

On the darker side of electoral choice, there is also a range of ways in which elites can manipulate electoral processes—"elections without democracy"—from repression to unfair rules to exclusion, coercion, corruption, and tutelage (Schedler 2002). Electoral systems also tend to affect how political parties and candidates shape their campaigns and strategies, with longstanding

[16] The types of cycles and the debates around them are described well by Reilly (2021).

concerns about democracy being vulnerable to extreme ethnic or nationalistic appeals by "ethnic entrepreneurs." There are debates over how to induce candidates to adopt appeals toward moderation (Reilly 2018) and ways the sequencing of elections can engender peace (Joshi, Melander, and Quinn 2017).

There is no perfect electoral system as each system involves some possible trade-offs or biases over such peace-supporting effects as legitimacy, inclusivity, accountability, or producing a cohesive, responsive government (Sartori 1968: 273). For example, the principle of majority rule is at odds with the principle of inclusivity. Systems that feature proportionality may contribute to peace but could also encourage volatility in the party system and ethnic polarization. Proportional representation can foster inclusion of different groups but could fragment the party system or may not mitigate the effects of majority domination (Bogaards 2013). However, whether any electoral system can lead to such desirable outcomes depends on many factors, such as the spatial distribution of votes, shifting party alignments and inter-party pacts, voting behavior, ballot design, and myriad other variables.

Without clear guidance and assessment of contextual conditions, policymakers risk mis-applying electoral recommendations to particular cases. Even with the very best assessment framework, data on voting behavior, or other tools such as polling, it is difficult to know how different electoral systems will perform through scenario analysis with available context-specific knowledge. There is therefore a tendency to adopt simplistic electoral systems to minimize these uncertainties. This may in turn lead to "missed opportunities" to move beyond an electoral processes merely becoming an "ethnic census."

Debate 4: Tilting Toward Partition? Decentralization and "Ethnic Federalism"

Fourth, ethnic federalism—as seen in Ethiopia, Nepal, Ukraine, and South Sudan—continues to be debated as both helpful and problematic. Under the right circumstances, territorially concentrated ethnic groups, particularly minority groups, can be accommodated through grants of autonomy. Agreements are reached between the rump government and the autonomous units over issues such as economic and foreign relations and regional commerce. Decisions on these limited issues are made jointly.

A first source of contention is over the possible configurations of ethnic federalism, each with pros, cons, and contextual requirements. Ethnic federations require more extensive interaction than confederations at the central government level, and questions around "federalism," or the allocation of powers and resources between the central and regional governments, is invariably a difficult and ongoing balancing act.[17] By contrast, a mixed or crosscutting federal approach may serve to mitigate conflict when groups are not territorially concentrated, or when the aim of federalism is to promote intragroup cleavages and foster alignments across groups. Mixed federations are appropriate when one or two ethnic groups are mobilized, aggrieved, and territorially concentrated but other groups are more integrated.[18] Noncommunal federalism can be especially appropriate when groups are not territorially concentrated or where significant minority communities live in various subunits. Dangers occur in mixed federations when special status is constitutionally conferred on one territory but not others (viz., Kashmir in India) and in noncommunal federations when some groups strive for greater territorial autonomy and such autonomy is not forthcoming.

A second unresolved question in the peacebuilding literature today revolves around efforts to "turn to the local" or to the urban, community, or small-scale territorial level to build sustainable peace, since local empowerment also entails controversies (Chandler 2017). Decentralization as a peacebuilding instrument has multiple and often contradictory implications.[19] The transformation of local governance following conflict can bring benefits to local communities, including greater political representation for diverse groups and an environment for innovation. Yet local government structures—whether formal state institutions or informal ones such as local peace committees—are vulnerable to capture and corruption.

The competing claims on the type of ethnic federalism and the degree of decentralization/local empowerment create openings for scholars to provide tailored analysis and recommendations to policy audiences. The choice of how to design ethnic federalism depends on a range of factors which can vary across countries and regions. Critical variables are the degree of economic interdependence, the structure of fiscal relations, the balance of dependency, and the size of the country—one argument which was made against ethnic federalism in Nepal (Bogati and Sisk 2018). As for "turning to local actors," careful, local studies of subnational conflict

[17] Other thorny issues in constructing ethnic federations are boundary delimitation; the structure of security; the containment of secessionist tendencies; relations between subunits and foreign governments and international organizations; disparities across region or state in the adjudication of law, language, and education policy; and—perhaps most important—the status of minorities and majorities within any given region. The escalation of civil war in Ethiopia in 2021 raised again deep questions about whether ethnic-federation solutions are sustainable or prone, by nature, to fissiparous tendencies.

[18] The territorially concentrated groups can be given special status or recognition while a nonethnic hue is preserved for the remainder of the polity.

[19] According to a cross-national study of local dynamics in eight countries undertaken by a consortium of US, Norwegian, Nepali, and South African researchers; see Suhrke and Sisk (2018).

dynamics are essential to identify effective interventions in highly complex local environments. They are also essential to identify appropriate actors who contribute to peace and practices that can avoid capture of governance by narrow interests (Suhrke and Sisk 2018).

Responsibly Coping with Competing Claims

Competing claims present responsible engagement challenges for both scholars and policymakers. In the case of the unresolved debates over ethnic power sharing, policymakers must make difficult choices given their best understanding of the issues and options in institutional design. The field is divided between ethnicity-based models, or consociationalism (group-based power sharing), and the alternative of centripetalism that seeks to disincentivize ethnicity as the organizing principle of politics. Yet the Annan-led approach in Kenya in 2007–2008 attests that at least some policymakers have been able to navigate and reconcile these competing claims, to craft solutions fit for purpose in local contexts at critical moments (Figure 4.1). In Kenya's

Figure 4.1 Kenya's President Mwai Kibaki (L) and opposition leader Raila Odinga sign a power-sharing agreement on February 28, 2008 in Nairobi to end a bloody two-month political crisis as former UN Secretary-General and mediator Kofi Annan looks on.
Photo by Simon Maina/AFP via Getty Images.

subsequent constitutional reform and in subsequent electoral processes, the original, transitional ethnic bargain was supplanted by a more centripetalist approach – including decentralization to limit presidential power – that seems to have been successful in stemming the slide into permanent ethnic power sharing (Shilaho 2022).

I enumerate several options to help cope with the challenges of competing claims. These include conducting additional research or applying additional research methods to describe the underlying social, geospatial, and institutional design considerations to identify context-specific recommendations or provide information about "unknown" factors; highlighting the trade-offs and consequences of particular policy choices; using "strategic imagination" to envision best- and worst-case scenarios; and learning policymakers' frames for processing conflicting claims.

A first option is for future scholarship to contribute to clarifying the findings and policy implications of the four debates on power sharing that continue to vex policymakers. Scholars may conduct additional research on the domestic drivers or social demands for inclusive, and accountable and responsive, executive institutions; better electoral system design that captures more fully social diversity and the complexity of implementation in country-specific contexts; inclusivity at local-level institutions; and mechanisms to connect elite interactions with local-level social cohesion and peacebuilding.

More specifically, scholars can intervene to moderate the claims of the efficacy of off-the-shelf-models and research. They can also be cautious in making claims about the efficacy of political institutions when their application and operation is so highly context dependent. They can instead explain conditional and medium-range implications of power sharing theories and tools for sustaining peace—such as the sensitive choice of an electoral system—that are specific to particular contexts.[20] Large-N quantitative studies of power sharing such as those of Hartzell and Hodie (2015) may clearly identify generalized patterns and support broad arguments for moving toward more democratic, inclusive politics, but the findings might not fit all cases. While there may be competing claims globally, specific claims (e.g., on power sharing) are often derived from and apply well to particular cases. Scholars can do more tailored research on sub-elements or research on particular countries. They can also broaden their research methods to be able

[20] The UN's Common Country Assessment methodology may be a helpful source: https://www.un.org/esa/socdev/unpfii/documents/ccaundafs.pdf.

to speak more accurately to conditions, such as mixed-methods approaches that include detailed case studies to complement global statistical results.

As part of their pursuit of more context-specific findings, scholars can also help to better orient policymakers toward these findings and the original research that generates them. The publications of major international organizations characterize the peacebuilding and power sharing debates for policy-oriented readers, including the UN–World Bank *Pathways to Peace* report and International IDEA's *Global State of Democracy* report. But their stylizations, summaries, and integrated knowledge from the literature can sometimes do more to obscure the original scholarly ideas they rely on or more nuanced source information from which they are derived. When helpful for decision-making, scholars can make these ideas and sources more accessible and transparent.

Because policy outcomes are often highly dependent on unknown factors, responsible scholars should also provide assistance with identifying possible influential contingent factors. They can then recognize and highlight the trade-offs of different policy options based on available research. Relatedly, they can exercise "strategic imagination" to infer how institutional arrangements designed to promote cooperation may play out if implemented. For example, List PR could lead to party proliferation, or it could shift degrees of ethnic or political identification. Scholars can help policymakers envision what are the possible best- or worst-case scenarios, given uncertainty. Despite decades of scholarly advocacy about what is "best," there are no one-size-fits-all or "holy-grail" solutions to the power-sharing debates. The hope is that scholars have at least identified options and assessed their advantages and disadvantages.

To succeed in carrying out the aforementioned guidance, scholars should learn how policymakers process competing claims. In the case of ethnic power sharing, when scholars disagree on fundamental approaches there may be limited visibility into how the competing claims are processed by policymakers. Indeed, knowledge about policymaker preferences is difficult to access given the secrecy of such processes, and it is rare to have detailed insights into them as in the case of the 2007–2008 Kenya mediation. However, there are a few identifiable modes of behavior. First, policymakers may often revert to their base preferences when presented with scholarly dissonance, simple models, sui generis case studies, and insufficiently contingent generalizations. They may thus advocate solutions based less on research and more on their own (often, country-specific) career-specific experiences. Examples of this rejection of research include the efforts by NATO to recommend PR in Afghanistan, advocacy of caucus-based primaries in Iraq, or advocacy for or rejection of referendums to resolve sovereignty disputes, as in Timor-Leste.[21]

Second, policymakers undertaking delicate mediation efforts tend to prioritize risk aversion and opt for solutions that may be suboptimal, such as the lesser-of-evils "ethnic power sharing" pact to stop the killing. They do not necessarily select what is "best" for ameliorating a conflict, but rather what seems to present the least risk of exacerbating it. Thus, for policymakers clinching even a minimally inclusive political settlement in protracted peace negotiations is a stronger imperative than the particulars of democratic institution design (especially given the uncertainty over whether such alternatives are better in the long run). Yet the policymaker imperative is strongly at odds with what scholarship may deem most effective for sustaining peace in the longer term. They may inadvertently lay the groundwork for an unsustainable peace over time by crafting political settlements that seal social relations in a particular moment in time. This might require accepting and even advocating ethnic power sharing—as various UN mediators have done recurringly in Cyprus—even if research has pointed to its problems and pitfalls, as an imperative to secure peace.

Third, even when policymakers broadly agree on the "best" solution—for example, the trend toward a mixed PR and constituency electoral system for a particular context—there are dilemmas of adoption by local protagonists. It is up to policymakers to craft preferences for tailored, on-the-table peace plans that can align with the preferences and bargaining power of local protagonists, and policymakers may espouse options which seem to be the most adoptable by them. In Bosnia, for example, there have been longstanding efforts by the OSCE to revisit the Dayton-era electoral system to break the grip of nationalist parties.[22] But these actors are uncertain or fail to see a direct or specific benefit to themselves in the adoption of alternative systems (since they are inherently more complicated and may be more vulnerable to electoral fraud or unexpected outcomes), which may tip policymakers away from advocating for them. Scholars can act more responsibly by identifying the deleterious effects of doing nothing in the face of uncertainty. They can articulate best-fit alternatives and advocate for institutional-form processes that may lead to their adoption.

[21] On Afghanistan, see Larson and Ramsbotham (2018): 119–122; on Iraq, see Dobbins et al. (2009); on Timor-Leste, see Fan (2008).

[22] However, there has been little effort on the part of local protagonists to significantly reform the electoral who presently benefit from the existing system (PR list); candidates in Bosnia must register their ethnicity in electoral processes. For an analysis of the 2022 elections and the continuing central role of ethnicity in the Bosnian electoral process, see Keegan (2022).

What do these behavioral tendencies of policymakers—reverting to defaults and accessible models, risk aversion, and prioritizing local actor preferences—imply for responsible engagement? First, because these preferences and incentives can produce suboptimal outcomes, policymakers themselves also shoulder an ethical burden and have key roles to play. When policymakers are hemmed in by contradicting academic advice, recalcitrant local actors, and even their own biases, they should work to become more self-reflective about these pressures. They may opt to avoid default solutions of ethnic bargains and instead work toward a sequenced transition of political change—toward inclusive politics that seeks to build broader social cohesion across identity lines—as in Kofi Annan's mediation in Kenya. Second, scholars should also be aware of the full range of design options and case-specific analysis and, as appropriate, consider how to nudge policymakers away from default modes or to broaden their views of solutions.

Conclusions

Social scientists encounter competing claims practically every time they conduct a literature review, in part because an imperative of scholarship is to make generalizable arguments for causal relationships. The gaps and puzzles that arise from such contention is part of what drives the process of research and discovery. But it also makes the at-times simple conception of "bridging the gap" and sharing research results with wider policy and public audiences more complex than it first appears. As I have argued, competing claims can, on some occasions, offer overly simplistic models to policymakers and give them license to select policies to which they were already committed, which can result in suboptimal outcomes. This is precisely the dynamic uncovered in the seemingly straightforward solution to ethnic-based conflicts of endorsing ethnic power-sharing arrangements as a means of cementing peace.

Upon closer inspection, the appearance of competing claims can also sometimes be specious. Sometimes competing research findings and policy claims can appear to be contrasting or competing at one level of analysis or in the aggregate, but findings may be conditional on context and there may be clearer answers for more specific levels of analysis. This is perhaps most evident in the broader set of issues to which the power-sharing debate speaks, namely that of advancing and deepening inclusivity more broadly. Even with competition among high-level claims, it can still be possible to offer clear advice related to decisions about conditions or types of cases.

Policy-engaged scholars working in areas of epistemic contention have a duty to clarify competing claims and lay out options and trade-offs as they offer advice, avoiding blanket advocacy of stylized models. They can conduct more research to better adjudicate claims or trace the possible consequences of decisions while recognizing the imperatives and uncertainties that policymakers face. Scholars then face a choice of how to engage. One path is to advocate for a single side of a debate—one they may implicitly favor—perhaps ignoring evidence. A more responsible and preferable option is to serve as an honest broker, fairly assessing literature, offering a range of options, and providing the best overall advice, even if counter to their personal research agenda or prior contributions.

References

Alihodžić, Sead and Nicholas Matatu, with Oliver Joseph and Katrin Lewis (2019) "Timing and Sequencing of Transitional Elections." International IDEA Policy Paper No. 18, April 9, 2019. https://www.idea.int/publications/catalogue/timing-and-sequencing-transitional-elections.

Andrés, Megan Neill (2021) "Power Sharing in Post-War Divided Societies: The Failure of Consociationalism in Bosnia-Herzegovina." *Strife Journal* 15(16): 143–152.

Aoun, Joy and Marie-Joëlle Zahar (2017) "Lebanon: Confessionalism, Consocialtionalism, and Social Cohesion." In *Peacebuilding in Deeply Divided Societies: Toward Social Cohesion?* ed. Fletcher D. Cox and Timothy D. Sisk. London: Routledge, pp. 103–136.

Associated Press (2020) "Sri Lankan Urges Limit on Minority Political Power." January 3, 2020.

Berry, Marie (2018) *War, Women, and Power: From Violence to Mobilization in Rwanda and Bosnia-Herzegovina*. New York: Cambridge University Press.

Bogaards, Matthijsn (2019) "Consociationalism versus Centripetalism: Friend or Foes." *Swiss Political Science Review* 25(4): 519–537.

Bogaards, Matthijs (2013) "The Choice for Proportional Representation: Electoral System Design in Peace Agreements." *Civil Wars* 15: 71–87.

Bogati, Subindra and Timothy D. Sisk (2018) "The Elusive 'New Nepal': Democratization, Ethnic Politics in a Plural Society," for the project "Forging Resilient Social Contracts." socialcontracts4forpeacee.org.

Chandler, David (2017) *Peacebuilding: The Twenty Years' Crisis, 1997–2017*. Cham: Palgrave Macmillan.

Cohen, Roger (2008) "How Kofi Annan Rescued Kenya." *The New York Review of Books*, August 14.

Coker, Margaret and Falih Hassan (2018) "ISIS is Weakened, but Iraq Election Could Unravel Hard-won Stability." *The New York Times*, January 30.

DeVotta, Neil (2021) "Sri Lanka: The Return to Ethnocracy." *Journal of Democracy* 32(1): 96–110.

Dobbins, James, Seth G. Jones, Benjamin Runkle, and Siddharth Mohandas (2009) *Occupying Iraq: A History of the Coalition Provisional Authority*. Santa Monica: RAND Corporation.

Fan, Hua (2008) "The Missing Link between Self-Determination and Democracy: Timor Leste." *Northwestern Journal of International Human Rights* 6(1): 176–195.

Harris, Peter and Ben Reilly, eds. (1998) *Democracy and Deep-Rooted Conflict: Options for Negotiators*. Stockholm: International IDEA.

Hartzell, Caroline and Matthew Hodie (2007) *Crafting Peace: Power-Sharing Institutions and the Negotiated Settlement of Civil Wars*. University Park, PA: Pennsylvania State University Press.

Hartzell, Caroline and Matthew Hodie (2015) "The Art of the Possible: Power Sharing and Post-Civil War Democracy." *World Politics* 7(1): 37–71.

Hellműller, Sara and Marie-Joëlle Zahar (2018) "Against the Odds: Civil Society in the Intra-Syrian Talks." International Peace Institute Issue Brief, March 2018. https://www.ipinst.org/wp-content/uploads/2018/03/1803_Against-the-Odds.pdf.

Holbrooke, Richard (1998) *To End a War*. New York: Random House.

Horowitz, Donald (1985) *Ethnic Groups in Conflict*. Berkeley: University of California Press.

Horowitz, Donald L. (2014) "Ethnic Power Sharing: Three Big Problems." *Journal of Democracy* 25(2): 5–20.

Howard, Lise (2019) "Assessing the Effectiveness of the UN Mission in the Central African Republic." *IPI Observatory*, November 21, https://theglobalobservatory.org/2019/11/assessing-effectiveness-un-mission-central-african-republic/.

Human Rights Watch (1995) *Playing the 'Communal Card:' Communal Violence and Human Rights*. New York: Human Rights Watch.

International IDEA (2017) "Inclusive Peacebuilding in Conflict-Affected States: Designing for Democracy's Resilience." In *The Global State of Democracy: Exploring Democracy's Resilience*, ed. Timothy D. Sisk. Stockholm: International IDEA.

International IDEA (2017) "Migration, Social Polarization, Citizenship, and Multiculturalism." In *The Global State of Democracy: Exploring Democracy's Resilience*. Stockholm: International IDEA.

Jarstad, Anna (2008) "Power Sharing: Former Enemies in Joint Government." In *From War to Democracy: Dilemmas of Peacebuilding*, ed. Anna K. Jarstad and Timothy D. Sisk. Cambridge: Cambridge University Press, pp. 105–133.

Joshi, Madhav, Eric Melander, and Jason Michael Quinn (2017) "Sequencing the Peace: How the Order of Post-Agreement Implementation Can Reduce the Destabilizing Effects of Post-Accord Elections." *Journal of Conflict Resolution* 61(1): 4–28.

Keegan, Jessica (2022) "Bosnia and Herzegovina's Extremely Complicated Electoral System Unpacked." International Republican Institute, October 1. https://www.iri.org/news/bosnia-and-herzegovinas-extremely-complicated-electoral-system-unpacked/.

Larson, Anna and Alexander Ramsbotham (2018) "Incremental Peace in Afghanistan." *ACCORD* 27: 119–122. https://www.c-r.org/accord/afghanistan/inclusive-politics-path-peace.

Lijphart, Arend (1977) *Democracy in Plural Societies.* New Haven: Yale University Press.

Lijphart, Arend (2004) "Constitutional Design for Divided Societies." *Journal of Democracy* 15(2): 96–109.

Lindenmeyer, Elisabeth and Josie Lianna Kaye (2009) "A Choice for Peace? The Story of Forty-one Days of Mediation in Kenya." International Peace Institute, August 2009. https://peacemaker.un.org/sites/peacemaker.un.org/files/KenyaMediation_IPI2009.pdf.

Linz, Juan (1990) "The Perils of Presidentialism." *Journal of Democracy* 1(1): 51–69.

McCulloch, Allison (2012) "Does Moderation Pay? Centripetalism in Deeply Divided Societies." *Ethnopolitics* 12(1): 111–132.

McCulloch, Allison and John McGarry, eds. (2017) *Power-Sharing: Empirical and Normative Challenges.* Abingdon: Routledge.

Mehler, Andreas (2009) "Peace and Power Sharing in Africa: A Not so Obvious Relationship." *African Affairs* 108(432): 453–473.

Minority Rights Group (2021) "Peoples under Threat 2021: Authoritarianism, London: Minority Rights Group International." December 16, 2021, at https://minorityrights.org/publications/put-2021/.

Nahan Calfat, Natalia (2018) "The Frailties of Lebanese Democracy: Outand Power Sharing in Africacomes and Limits of the Confessional Framework." *Contexto International* 40(2): 269–292

Nilsson, Desirée (2012) "Anchoring the Peace: Civil Society Actors in Peace Accords and Durable Peace." *International Interactions* 38(2): 243–266

Ould-Mohamedou, Mahmoud and Timothy D. Sisk, eds. (2017) *Democratization in the 21st Century: Reviving Transitology.* Abindgon: Routledge.

Paffenholz, Thania ed. (2009) *Civil Society & Peacebuilding: A Critical Assessment.* Boulder: Lynne Rienner.

Pérez-Liñán, Aníbal, Nicolás Schmidt, and Daniela Vairo (2019) "Presidential Hegemony and Democratic Backsliding in Latin America, 1925–2016." *Democratization* 26(4): 606–625.

Reich, Robert (2017) "First 100 Days: Donald Trump and the Degradation of the Presidency." April 23. http://robertreich.org/post/159905075775.

Reilly, Benjamin (2021) "Cross-ethnic Voting: An Index of Centripetalist Systems." *Government and Opposition* 56(3): 465–484.

Reilly, Benjamin (2018) "Centripetalism and Electoral Moderation in Established Democracies." *Nationalism and Ethnic Politics* 24(2): 201–221.

Reilly, Benjamin and Per Nordlund, eds. (2008) *Political Parties in Conflict-Prone Societies.* Tokyo: United Nations University Press.

Roeder, Philip and Donald Rothchild, eds. (2005) *Sustainable Peace: Power and Democracy after Civil War.* Ithaca, New York: Cornell University Press.

Sartori, Giovanni (1968) "Political Development and Political Engineering." In *Public Policy*, ed. J. D. Montgomery and A. O. Hirschman. Cambridge: Harvard University Press, pp. 261–298.

Schedler, Andreas (2002) "The Menu of Manipulation." *Journal of Democracy* 13(2): 36–50.

Shilaho, Westen K. (2015) "Third Time Lucky? Devolution and State Restructure under Kenya's 2010 Constitutional Dispensation." In *African State Governance*, ed. A.C. LeVan, J.O. Fashagba, and E.R. McMahon. London: Palgrave MacMillan.

Stewart, Francis (2000) "Crisis Prevention: Tackling Horizontal Inequalities." *Oxford Development Studies* 28(3): 245–262.

Suhrke, Astri and Timothy D. Sisk (2018) "Innovations in Peacebuilding: International Norms and Local Dynamics in Conflict-affected Countries." Final Report. Josef Korbel School of International Studies, March 2018.

Tamaru, Nananko and Marie O'Reilly (2018) "How Women Influence Constitution Making After Conflict and Unrest." Inclusive Security Research Report, January. https://www.inclusivesecurity.org/wp-content/uploads/2018/02/How-Women-Influence-Constitution-Making.pdf.

United Nations Development Program (2020) "Strengthening Social Cohesion: Conceptual Framing and Programming Implications." https://www.undp.org/publications/strengthening-social-cohesion-conceptual-framing-and-programming-implications.

Wimmer, Andreas, ed. (2004) *Facing Ethnic Conflicts: Toward a New Realism.* Lanham: Rowman and Littlefield.

Wolff, Stefan (2004) "Settling an Ethnic Conflict through Power Sharing: South Tyrol." In *Managing and Settling Ethnic Conflicts: Perspectives on Successes and Failures in Europe, Africa*, and Asia, ed. Ulrich Schneckener and Stefan Wolff. London: Palgrave MacMillan: 57–76.

5

Partnered Policy Engagement

Positionality and the Ethics of Collaborating with Civil Society Organizations

Oliver Kaplan

> For every person who seeks to innovate, bring your personal perspective reflecting who you are and where you come from to shape a future in which every person can thrive.
>
> **–Arati Prabhakar, Director, White House Office of Science and Technology Policy (2023)**

> It is difficult to know oneself, but it isn't easy to paint oneself either.
>
> **–Vincent van Gogh**

What happens when a research subject becomes a research and briefing partner? In 2017, I was contacted by Peace Direct, a peacebuilding NGO that works to connect local peacebuilders with policymakers, to contribute to a policy report on community-based atrocities prevention.[1] I invited Cristina Serna, a local peacebuilder I knew from Colombia, to partner with me in the endeavor. We co-facilitated an online forum and drafted a chapter for the report (Kaplan and Serna 2018). We then shared our findings together—plus her experiences and my research—at in-person briefings with NGOs and policymakers in the US. I came to refer to what we did as "partnered engagement." This chapter explores how scholars can empower—rather than

[1] My partner in engagement, Cristina Serna, was consulted throughout the process of drafting this chapter, and her input has been incorporated herein. My gratitude goes to Sapna Considine, Dimitri Kotsiras, Bridget Moix, Kelsey Shantz, and Jennifer Smyser for the opportunity to participate in the engagement events and providing helpful inputs for this analysis. I am grateful to Johanna Amaya-Panche, Gina Bateson, Rachel Epstein, Cullen Hendrix, and Dan Mahanty for helpful comments. The views expressed herein are the author's and do not represent positions of Peace Direct or the Stanley Center for Peace and Security. Punctuation of the epigraph quote by Prabhakar was edited for clarity.

Oliver Kaplan, *Partnered Policy Engagement.* In: *Speaking Science to Power.* Edited by: Rachel A. Epstein and Oliver Kaplan, Oxford University Press. © Oliver Kaplan (2024). DOI: 10.1093/oso/9780198875185.003.0005

marginalize—stakeholders who are the subjects of research during the policy engagement process. It raises the question of "Who has a seat at the table?" and grapples with how to ensure that a wider range of voices reaches the ears of policymakers.

Partnered engagement is a relatively new and still infrequent form of policy engagement generally consisting of collaboration between academics, local peacebuilders, and NGOs in the dissemination of research findings and policy implications. This chapter specifically focuses on the sharing of the successes and failures of community-based atrocities prevention strategies with other NGOs and government policymakers. This approach differs from traditional research processes, where local partners may primarily be contributors of raw information for academic analyses but are left out of the policy engagement process (and whose views may therefore end up filtered by researchers). By contrast, the partnered approach directly includes these voices in the engagement.

Although I focus here on a process with civil society partners, this form of policy engagement may arise across a broader set of engagement scenarios. One example is the possibility of conducting joint presentations of research findings by academics and bureaucrats (or academics and activists) to higher-level policymakers. Berman et al. (2018) report instances of academics partnering with subordinate military officials to brief research findings to higher-ranking commanders. In the case of the Roots of Restraint project supported by the International Committee of the Red Cross (ICRC), academics partnered with ICRC staff to develop findings for the rest of the organization, its leadership, and the broader humanitarian sector.[2] The Center for Civilians in Conflict (CIVIC) is another NGO that directly partners with local leaders and activists and strives to engage with them on issues of civilian protection at every step of their programmatic, research, and advocacy processes with national and international policymakers (according to CIVIC staff, personal correspondence; CIVIC 2020, 2021). These reflections could provide guidance for these types of situations as well.[3]

My partnered engagement experience began with Peace Direct's invitation to contribute to a policy report titled *Atrocity Prevention and Peace Building*.[4] I agreed, but given my assigned topic of "Local Self-protection Strategies in Atrocity Prevention," I felt it would be artificial

[2] The International Committee of the Red Cross, "What restrains armed actors from committing atrocities?" May 17, 2016, https://www.youtube.com/watch?v=FgTMp8EaZHI.

[3] However, the power differential between academics and engagement partners in the circumstances may differ from that of partnering with local peacebuilders.

[4] Peace Direct is an international NGO that works with local people to stop violent conflict and build sustainable peace. It supports local organizations in eleven countries around the world and advocates for

for me alone to simply relay the voices of local actors. I insisted that this be in partnership with a local peacebuilder. I contacted Cristina Serna, then president of the Peasant Workers Association of the Carare River of Colombia (ATCC, by its initials in Spanish), to invite her to become my coauthor.[5]

Cristina and I met during my first research visit to her community ten years earlier. In my book, *Resisting War*, and in other writings, I analyze their community-protection approaches in hopes of sharing them with both English and Spanish-speaking audiences (Kaplan 2013, 2017). Cristina and I kept in touch, and I went back to La India, her community's main village, to share my research with them (Kaplan 2020). This has been a fruitful example of ongoing, two-way, engaged research and advocacy. But with our new, partnered engagement we would be taking our action-oriented research from the field to the briefing room. With this shift in setting came new questions of positionality, such as those first voiced by feminist geographers (e.g., England 1994; Rose 1997): What were our different social positions, and how were our positions viewed by others?

These questions shaped the ethics of our interaction since we came from very different backgrounds. I would like to say I was explicitly aware of my positionality and motivations in policy engagements and took these questions on prior to our shared experience, but that is not the case. I believe I was implicitly aware of Cristina's and my differences and sought to elevate my partner, but upon reflection I was not fully aware of our positionalities or what they implied for our policy conversations. I did not realize what a change the setting would be for Cristina from her daily experience, or how relatively familiar it would be for me. Positionalities may only be most apparent once an engagement opportunity has commenced in person, with both partners present before a previously unfamiliar audience, and then the partners can adapt. Still, that is no excuse for not initially assessing these questions beforehand as well. Despite these unanticipated challenges, our cooperation in this instance could generally be considered successful.

Our partnered engagement also presented pathways to address several deeper ethical issues that can arise during policy engagement activities. As

locally led approaches to preventing conflict and building peace. Its *Local Voices for Peace* project organized an online consultation, held in November and December 2017, looking at the relationship between peacebuilding and atrocities.

[5] The ATCC was founded in 1987 to gain autonomy from multiple threats by guerrilla (FARC and ELN), paramilitary, and state actors, and is one of Colombia's earliest local peace organizations. In its "areas of influence" in North-central Colombia it represents and defends thirty-two villages. It sits in a steamy, forested valley crisscrossed by river tributaries, about twelve hours by bus from the capital, Bogotá. From there it is even farther by foot, donkey, or motor-canoe to reach outlying villages. As residents proudly announce, they were the winners of the Right-Livelihood award (Alternative Nobel for Peace) in 1991.

a more direct form of communication—less of a game of "telephone," where insights may be lost in translation via the academic researcher—the approach could provide more accurate and impactful information to policymakers (e.g., a clearer picture of George's [1994] "image of the adversary" or the actors involved in a strategic situation; or perhaps representing a kind of temporary "embedded capital" in the parlance of Nye [2008]). The local partner's presence is a costly form of engagement that may attract greater policymaker attention, and the partner can convey more genuine stories, specific examples, and nuances based on their experiences. At the same time, local partners may incur risks in some cases by engaging with national or foreign governments, whether directly or when others convey their experiences. Involving the partners in the engagement process provides them with greater control over how their views and findings generated from their lived circumstances are communicated to policy audiences.

Ethical partnered engagements with local actors can contribute to the efforts of policy-focused NGOs spanning the community development and international policy realms as well as government policymakers seeking to improve foreign assistance policies. For my part, I saw the opportunity as a rare chance to foreground local voices while also drawing on my past research to add occasional interpretation to such voices. In outlining this engagement approach, I first describe the process of our interactions with policymakers. Second, I consider the moral imperatives, benefits, and positionalities of partnered engagement, as well as the possible ethical risks. Third, I discuss the necessity and usefulness of evaluating the results of partnered engagements—getting feedback on the "black box" of policymaking. I conclude with some ideas for improving partnered engagement in future applications.

Our Process of Partnered Policy Engagement

Partnerships between those in and out of the ivory tower or between international and local actors are frequent, despite stereotypes to the contrary. For instance, Participatory-Action Research involves communities in conceiving, designing, and implementing research for the benefit of communities (Chevalier and Buckles 2013). Unarmed accompaniment of local human rights activists may be carried out on the ground by international human rights activists, and even at times by professors or intellectuals, to provide protection through the means of mere physical presence, monitoring and publicizing rights violations, or advising (Mahoney and Eguren 1997).

Academics and practitioners or activists may also partner for public discussion panels. Rarer, to my knowledge, is the focused researcher–practitioner teamwork for communicating the experiential and research-based policy insights described herein.

Co-facilitating Discussion of Atrocities Prevention with Local Peacebuilders

Cristina's and my first step was to gather the perspectives of ninety peacebuilders, academics, and professionals from around the world through an online Peace Direct "consultation" that lasted twenty-four hours. As conversation facilitators, we first jointly crafted a guiding statement for the consultation, along with the initial themes and questions to which the participants would respond. Cristina provided an invaluable reality and legitimacy check for the questions that I intuited were important but of which I could not be certain. She was also able to include insights from her experience and her community. Our session was a great success in that it included diverse participants and perspectives from Bangladesh, Burundi, Nigeria, Sri Lanka, and other countries on community strategies for confronting atrocities.

Our analysis distilled points of consensus among the participants and key policy implications. The participants reaffirmed the existence and value of both *formal and informal* local peace organizations (the ATCC itself is a formal peace organization, but it also has informal constituent village councils). There was also consensus about the value of three strategies or approaches for community self-protection, including early warning, local mediation and reconciliation, and skilled local leaders.[6] As Cristina noted (and as I have studied), the ATCC has its own early warning system and a group of men and women conciliators who mediate disputes among residents and with armed actors to protect accused civilian informants and collaborators.

Managing Our Policy Brief Coauthorship

Even with our strong relationship, coauthoring a policy brief with an activist was rewarding but not without obstacles. I saw the consultation and report

[6] As a note of caution, there may be an *accessibility bias* inherent in the participants' comments, since they may have focused on default strategies, or the most well-known and publicized strategies that are part of the existing international peacebuilding vocabulary, such as community early warning systems.

section as a way to directly involve a local activist in the dissemination process. If anything, it might at least lessen the risk of just "taking" information and insights from a community during the normal research process and then analyzing and relaying it to broader academic and policy communities—all to the benefit of my academic career (what some have termed "extractivism"). I felt this could amplify an activist's voice and provide a chance for them to build their international networks as well as benefit from an opportunity for travel and cultural exchange. As such, I wanted to make sure that Cristina was completely involved in the preparatory steps for the consultation, as well as in the moderation of the dialogue and the drafting of our analysis and policy brief. This was not as simple a task as it might first seem.

Since the activity was conducted in English and Cristina had greater constraints on her time (and due to my academic and writing background), I managed the collaboration process. In the preparation phase, I was sure to include Cristina's input on the themes and questions for the consultation. Then, during the consultation, I devised a procedure to ensure that she could both follow the English-based online conversation and provide her own feedback from afar in Colombia. Cristina was visiting isolated villages on the days of the consultation and did not have reliable internet connections in the *campo* (countryside) of her isolated corner of Colombia. To manage this challenge, as participants posted comments I translated them from English to Spanish using Google Translate and sent the text to Cristina through a WhatsApp chat, and then back-translated and posted her responses to the online forum.

An additional and unforeseen challenge for collaboration stemmed from Cristina's position as the leader of her community at the time. Because of her role and attendant responsibilities, Cristina had to conciliate threats against residents of her community *during the very days of the consultation* (and in addition to her regular employment). These threats were likely from criminal bands that emerged in Colombia in the wake of the demobilization of the AUC paramilitaries in the mid-2000s, including the ATCC region. This of course limited her availability and attention, but it did seem somewhat more urgent than academic analysis and forum moderation. This meant, however, that I had to balance involving her in the process with not burdening her with excessive demands on her time to review material or provide input.

At the post-consultation write-up phase, I solicited her input for the content of the report section and took it upon myself to write a draft. I then circulated it to her (in Spanish), verbally summarized the key claims, incorporated her feedback, and submitted it with her approval.

Briefing Policymakers

To directly engage with international policymakers and share the results of our document, Cristina came to Washington, DC, where we met for dinners with policymakers sponsored by the Stanley Center for Peace and Security and had additional side meetings (Figures 5.1 and 5.2). Participants included officials from international organizations, such as UN agencies, as well as NGO staffers and US government officials (e.g., State Department and US Institute of Peace, among others). While I was interpreting for her from Spanish to English at the events, I also sought to keep her voice at the forefront. For the ATCC community, the NGO and policymaker connections proved highly valuable in being able to share its lessons with a broader audience. The experience also helped Cristina, the local peacebuilder, gain at least temporary support from new Colombian and international actors for her community's violence prevention work. The participation of a local voice and face in the dialogues also gave more weight to the previous scholarly research

Figure 5.1 Cristina Serna leads a community planning meeting during the anniversary celebration of the founding of the ATCC, La India, Santander, Colombia, 2013.
Photo: Oliver Kaplan

Figure 5.2 Oliver Kaplan and Cristina Serna at the US Institute of Peace in Washington, DC, 2018.
Photo: Oliver Kaplan

findings by humanizing them.[7] More broadly, I felt a sense of responsibility to support and coach her through this foreign process in a foreign country, including with her travel and other logistics while in the US.

Reflections on the Ethics of Partnering

Along with the dissemination successes, the partnership process also gave rise to several unexpected challenges, risks, and possible negative effects.

[7] CIVIC similarly sees their work as more effective and sustainable when communities directly participate, which generates greater social cohesion and confidence, adaptability to future threats, and influence with armed actors (CIVIC 2020, 2021). For example, CIVIC's analysis of the application of Iraq's Compensation Law was based on feedback from local partners.

Benefits and Risks for Local Partners

For the local peacebuilders, the visit opened up previously unimaginable access to policymakers and officials, as well as the prospect of broadening their professional networks. The trip was also a unique travel opportunity for Cristina, and her involvement with the report was a way to provide her with a small amount of remuneration for her efforts (as Peace Direct paid participants a fee).

Alongside the possible advantages that engagement activities may offer local participants, they can also impose hardships. For instance, logistical issues were a challenge in this case. Cristina required a visa and a passport, which were time-consuming for her to acquire and were issued late, causing her to miss an initial briefing event in New York (and requiring me to convey comments on her behalf in that setting). Organizing the participation and travel also posed opportunity costs and potential hardship on the local peacebuilder by reducing the available time for her daily peacebuilding work and economic activities. There was a trade-off between the impact of her participation and face-time with policymakers and her time away from *actually building peace.*[8] To limit these possible difficulties, conveners should analyze such potential burdens early on and consider ways to offset them with additional resources or support as needed.

The process of engagement, while seeking to include new local voices, can also elevate some countries, communities, and individuals over others. In this way, some individuals may be relatively more privileged in being able to benefit from the opportunity to travel internationally and engage in high-level meetings. The emphasis on participation of English speakers in the consultation and report-authoring ended up being exclusionary in some ways.[9] The African peacebuilder participants who joined the events could speak and present in English, but Cristina was dependent on my (hopefully adequate) interpretation.

It was certainly more convenient to have English as the *lingua franca* of the convening because translation is complicated and expensive. However,

[8] While she saw me as a "good host" and felt well-informed ahead of time about the conditions she would encounter in her engagement, the experience moved Cristina a little outside of her comfort zone. For instance, even small things such as not eating her customary foods could present difficulties (as she said, "I'd prefer more Latin food—there's not even *changua* [a Colombian soup] here!").

[9] According to Peace Direct's assessment of the local peacebuilders at the engagements, "they tend to be from more urban areas that have access to the internet or can be easily accessible during country visits by policymakers and practitioners. This excludes many people that live in rural or hard to reach parts of the country from having a voice in what could be the best approach to resolving conflicts in their nation or region."

individuals who, as non-English speakers, may already be disadvantaged in their access to international actors and resources are thereby further excluded from such engagement activities. Yet non-English speakers have valuable insights to offer about locations and issues that English-speaking policymakers frequently have less access to (relative to English-speaking regions)—a selection bias that limits the views available to primarily those from the "globalized" locals. To manage the issue of language-related exclusion, conveners should consider innovative technologies, partnering with multi-lingual individuals, or making additional investments in translation services to broaden the profiles of engagement participants.

Finally, the selection of community representatives for engagement activities has the potential to cause rifts and jealousies within communities. Participation can confer benefits such as travel, access, prestige, or increased power within one's community. Some convening organizations may be inclined to repeatedly invite the same individuals for engagement activities because of familiarity, seeking to avoid the set-up costs of making initial arrangements (such as applying for visas, etc.). While involving repeat participants can facilitate building sustained relationships, there can be a trade-off of concentrated participation benefits for a few individuals and fewer new voices. Conveners should be attuned to the fairness of the selection of participants since invitations themselves could exacerbate cleavages among the very groups or communities that policymakers would seek to assist. They should therefore consider expending resources to invite additional participants or integrating additional forms of access or participation—whether virtual or via another form—to include other individuals in the engagement activity.

In sum, involving local actors in research dissemination and policy engagement activities can create both rewards to the various parties involved as well as ethical hazards that must be taken seriously. These risks and rewards may vary across policy engagement contexts. Scholars and conveners should continue to work to identify these potential hazards and devise pathways to address them.

Identifying, Managing, and Leveraging Positionality during Partnered Engagement

Our partnered process also put the issue of my positionality as a researcher and engagement partner front and center. Accounting for one's positionality is the act of recognizing one's relative social position and how others may conceive of it (England 1994; Bourke 2014). Positions of prestige and influence can be helpful—even critical—to leverage one's access and power to push for policy changes. However, Prabakar's (2023) epigraph quote also observes that everyone can draw on their unique identities to make contributions to science and policy. Everyone can identify and manage their positions and positionality in the context of the various relationships of the engagement process.

In establishing our academic–peacebuilder partnership, I could not ignore our differing incentives and power imbalances, especially in a setting like Washington, DC. Cristina is an Afro-Colombian woman from a small rural community, with leadership skills but language limitations, and was on her first international trip. By contrast, I am a white man, working in a familiar country, city, setting, and language, and had the benefit of the (modest) prestige of being a PhD researcher—perhaps helpful to get in the door—plus some base level of legitimacy from past field experience. In this sense, I was more of an "insider" in this context, while Cristina was more of an "outsider" (Chavez 2008; Merriam et al. 2001). However, the demands of policymakers for Cristina's unique real-life experience plus my academic position proved to be a powerful combination: it boosted access while helping communicate policy implications in a structured, heartfelt, and ground-truthed way.[10]

Differences in positions can also come with some risks, however. As a "global north" researcher, I had an air of familiarity and ease of communication (or simply was more easily comprehensible), which could have overshadowed my partner. To "manage" positionality, it is therefore important to explicitly recognize and assess positions and what they imply in terms of how partners might uniquely interact with policymaker participants. This entails articulating what characteristics, access, advantages, and limitations each individual has, which can vary based on each person's professional and life experiences (and even from one type of policy audience to another). This raised several important questions about my role: Was I representing my own work and findings? Was I accurately interpreting what she said? Was I translating and clarifying what she said for greater comprehension and impact? Or was I drifting between these modes across the course of the engagement?

With at least a modest perception of our relative positions, we were able to leverage our teamwork for impact. We combined the more removed and abstract academic analysis with more detailed and practice-based insights. However, at times I did attempt to translate, clarify, and interpret Cristina's comments for policymakers based on my past experience with their general

[10] Additionally, as a researcher, I was aware that the act of engagement alongside a local community pitted my natural research ethos of seeking to remain impartial or unbiased against a policy advocacy role, possibly blurring the lines between objectivity and partiality.

decision-making frameworks and what kinds of insights, evidence, and examples would most likely be internalized. For example, in describing her community's activities, Cristina explained their mediation processes to resolve violent threats by armed actors, but without providing many specific details.

In recounting the ATCC's mediation, Cristina may have perhaps assumed people naturally understood her community's process, or maybe she was simply not used to explaining it in (analytical) detail. But I knew that it is the nuances of her community's mediation system that makes it so effective and adaptable, and that additional detail would be helpful for policymakers to potentially export it to other cases. So, like an interviewer who elicits stories from their interlocutors, I prompted her to mention key aspects I knew were important from my prior interactions and field visits with her and her neighbors. I encouraged her to discuss the sources and processes they use to gather information to investigate threats and the various ways they seek to maintain peaceful social norms, going beyond simply stating that these features of her community organization exist. I also occasionally added my more analytic perspective after her remarks.

Given the questions surrounding positionality, adhering to some guidelines or best practices can help to ensure ethical and equitable engagement. One suggestion is to apply Immanuel Kant's (1781) notion of a "categorical imperative": valuing something or someone as an end in themselves, and not using them as a means to achieve some other end.[11] In this context, this principle can guard against the local actor being "used." It implies that involving a local actor in an engagement process should be done with the primary aim of attaining some greater good, such as helping the local actor communicate their knowledge or needs, and helping policymakers to craft pareto-improving policies. It should not be undertaken for virtue-signaling, gaining credit or recognition, or (strictly) promoting ones' own aims or distinct research findings.[12]

The concern of disproportionate benefit on the part of the academics in these relationships should be kept in mind. However, it should not generally be considered an ethical breach if researchers derive some secondary benefit from interactions, so long as the local participant's interests are prioritized.

[11] "Some ends are merely instrumental, that is, they are sought only because they serve as 'means' towards further ends. Kant argues that the moral law must be aimed at an end that is not merely instrumental, but is rather an end *in itself*." *The Internet Encyclopedia of Philosophy* (IEP), https://www.iep.utm.edu/kantview/#SH5b.

[12] Indeed, I continue to worry that this is in fact what I ended up doing, or that this may be how the engagement experience or even this chapter may be perceived. I also believe, however, that it is these very worries that help keep researchers "honest" in their conduct vis-à-vis other actors.

From a Kantian perspective, these other reasons and benefits for academics to participate in partnered engagement are not amoral so long as the duty to the civil society actor or partner is the "operational reason" for our action (e.g., BBC 2014). In other words, the primary motivation of the interaction should be putting the partner's interests first. This helps ensure that the partners are not being treated as means to ends while not neglecting the possibility that some other benefits may accrue from the interaction.

In most circumstances, exploitation of the local actor is probably not an issue because it can be quite costly for a researcher to go out of their way to aid a local actor in engaging with policymakers (in terms of time if not also resources). As such, it would mainly make sense to participate in the engagement from a moral, altruistic perspective and not from some other cost–benefit calculation. Still, the categorical imperative frame remains a wise precautionary principle to abide by to avoid ethical hazards. The partner should also enter the engagement with full and informed consent based on knowledge of the possible benefits, costs, and risks, as far as they can be anticipated. To go beyond simply the minimization of harm or exploitation, researchers can further work to identify what additional assistance, if any, their engagement partners may require.

Assessing Engagement: An Ethical Imperative

Because partnered policy engagement offers many advantages but can also impose costs, it is advisable that its usefulness be checked and evaluated to make sure it provides added value for its various participants. Fortunately, both Peace Direct and the Stanley Center for Peace and Security, which sponsored the engagement dinner meetings, conducted assessments of their activities. There is much to learn from these exercises. The results indicate general expressions of gratitude for the interactions on the part of the local peacebuilders with whom I partnered (as well as from other activities). The policy participants also voiced positive responses and found the experiences and information communicated to be valuable for their future work, though some also reported that their views had previously already been informed by local perspectives.

For a better sense of the evaluation process and findings, I briefly summarize and analyze the findings of the evaluation reports (Peace Direct 2019). Although these evaluations represent self-reflections on the part of the sponsor organizations and there are no comparisons with other forms of content delivery, they provide helpful initial insights. Peace Direct conducted surveys and seven-question interviews both before and after the engagements with several sets of local (LVP program) peacebuilders to probe how useful and impactful the visits were for the peacebuilders.[13] The Stanley Center also fielded a small follow-up survey of policy participants between April 26 and May 18, 2018 to understand "how much of local peacebuilding impacts or is used in their work."[14] Of a total of fifty-one participants in the roundtable events, a modest seventeen respondents completed the survey (nine [out of twenty-three] from Washington, DC, and eight [out of twenty-eight] from New York).

The results first indicate the utility and use of the engagement exercise. According to the Peace Direct responses, 86.5 percent of participants stated that perspectives from local peacebuilders are "a necessary and critical part of understanding conflict situations," compared to 13.3 percent of participants at baseline (an increase of 73.2 percent). Per the Stanley results, despite many missing responses, four-fifths of respondents (fourteen of seventeen) felt that the discussions helped them "reach new levels of depth or raise questions that need closer examination," and one participant lamented they "just wished we had a bit more time [at the event]." Still, there was also some "preaching to the choir," as many participants (thirteen of seventeen) felt that local expertise had already informed their decision-making (as one commented, it is "good reinforcement if not necessarily new information"). Thus, these exercises might perhaps have the greatest marginal impact with policy participants who are interested in learning from local actors but who have not previously enjoyed many such opportunities (perhaps individuals farther away from the organizing NGO's traditional policy networks). "Preaching to the choir" should also not necessarily be considered problematic, since it may still provide key policy advocates with additional "ammunition" (in the most peaceful sense!) to elevate new peacebuilding perspectives solutions.

Participants generally viewed the local peacebuilders' remarks as "validating" for them and could specifically recall points that had "stuck with

[13] Of respondents to the survey, 51.3 percent were either a part of an NGO as staff or a practitioner, while policymakers made up only 0.6 percent. Academics made up 29.7 percent of respondents, and there was one individual who identified themselves as a "Government Donor/development Official" (I was apparently a respondent to this PD survey).

[14] Questions included:

"What steps or actions have you taken that were motivated or inspired by ideas discussed at the policy salon?"

"What thoughts, ideas or comments from the policy salon, if any, have stuck with you? Will anything in particular inform your work going forward?"

"Has the event, your interactions at the event, or your interactions after the event influenced your opinions of local voices and their role in the decision-making process?"

"[Did] Discussions at the policy salon enable participants to reach new levels of depth or raise questions that need closer examination?"

them." In terms of substantive utility, they provided new insights on conflict dynamics, rumors and misinformation/freedom of information, and the role of the media in the perpetuation and prevention of violence. They also helped to "connect the dots between local and international efforts," "identify local partners," and design "more meaningful monitoring and evaluation." Some key feedback also overturned standing assumptions: one person came to realize that some peacebuilders feel that "the international community is not listening to or trusting local civil society organizations."

The interviews also point to challenges with traditional forms of policy engagement and the potential advantages of partnered engagement. Many claimed a main challenge is the "lack of trust between the US government and the local organizations." An anonymous participant suggested that the "biggest challenge in engaging with local communities are the [politician or community] gatekeepers who ... often give biased information." This can harm the credibility and legitimacy of local peacebuilders trying to engage with policymakers. This speaks to the utility of direct or partnered engagement, which may help ameliorate this problem. As a policymaker in the survey stated, "Every time a local peacebuilder is brought into the State Department for a visit, staff realize the valuable information that locals bring to the table."

The peacebuilders greatly valued the access afforded by the engagements and the related increase in their "visibility." One peacebuilder participant said, "The visit makes it possible to deliver our voice of peace to different decision-makers that otherwise wouldn't be possible." Another emphasized that the visit provided an entry to USAID that was not available before. Cristina was similarly able to meet with her country's embassy in Washington, DC, as well as with several US government agencies.

These analyses of the engagement were not able to take the final step of testing for their possible impacts for actually generating shifts in the ultimate outcome of changes in policies. According to Bridget Moix from Peace Direct, this is a key future goal for assessment: to identify "if the resources and actions are really shifting toward local peacebuilders and [if there is] more effective work to prevent violence and build sustainable peace" (personal correspondence).[15] As a preliminary signal, the Stanley study did show that at least two-thirds (twelve of seventeen) of respondents reported discussing their experience at the policy dinner with a colleague outside their organization. Moix also noted that the feedback from the evaluations was already driving

[15] It may not always be obvious how insights from participants can be directly connected to policy recommendations and outcomes. In some instances, the engagement may generate more general lessons and insights. This can make it difficult to evaluate the final step of the effects of these policy engagements.

change within her organization: “We have also been improving upon our own internal systems for peacebuilder visits and developing much more in depth safeguarding policies across all our work.” As an NGO staffer mused, “How to sustain the momentum post-visit is something we all struggle with.”

Conclusions and the Future of Partnered Engagement

Partnered engagement with NGOs and local actors is an exciting and impactful new approach for communicating research findings and practice-based knowledge with government and “elite” policymakers. It entails collaboration and strategizing among partners, and is not simply a case of giving separate, uncoordinated remarks side by side. This approach promises various benefits for the different actors involved, including new experiences and networking for civil society actors and successful brokering and advocacy roles for NGOs. Policymakers can also gain more legitimate and unique perspectives and lessons and academics benefit from broader dissemination of research findings. It can be a form of inclusion for partners, so that research relationships are sustained and not simply built around information “extraction.” It brings more seats to the policymaking table.

At the same time, there are also potential costs and risks, especially for the civil society actors who may be the relatively more vulnerable partner in the relationship. The potential costs and risks for the civil society actors (including diversion from daily activities and potential exclusion from participating) should be analyzed and accounted for by engagement organizers. This may inform when partnered engagement is worthwhile relative to traditional forms of engagement (e.g., briefings by scholars alone).

Partnered engagement acquires its persuasive power precisely from the different positions held by the different participants. Civil society actors bring influence because of their direct experiences, while academics can contribute research-based evidence and assist with interpreting, emphasizing, and explaining key insights brought by their engagement partners. However, the positionality of the participants must also be taken into account, since there could be differential benefits and vulnerabilities for each of them. Positionalities should therefore be explicitly identified, managed, and then leveraged for ethical policy impact. A guiding principle espoused for this endeavor is Kant’s categorical imperative, which helps ensure that the civil society participants’ interests are prioritized.

Quality assessment of partnered engagement activities is a final best practice. Indeed, partnered engagement is only clearly a win–win if the potential risks and benefits are measured and managed. If academics and NGOs are going to engage with local actors they should work to ensure such exercises are worthwhile (given their potential costs and risks, as well as the challenges to do it well and sustain impact). Assessments help verify this. If policymakers are going to benefit from the contributions of local actors, they should consider it a duty to respond to evaluations to both validate and improve the interactions (since civil society actors may travel for several days to provide them with a unique and enriching experience). Policymakers often juggle busy schedules, but so do civil society actors. Contributing to evaluations is the least policymakers can do to uphold their end of the engagement.

Assessment is also instructive for providing feedback to scholars about the policymaking "black box" and improving the impact of engagements. There should therefore be more evaluations with greater independence from implementing organizations. These analyses could compare traditional formats (briefings) with other approaches and check whether engagements go beyond simply "educating" and directly contribute to actual shifts in decisions or policies, either immediately or down the road.[16]

There remains much to learn about the practice and effects of partnered engagement, so collecting additional examples would be of great value, including examples from other policy domains besides peacebuilding. As the settings, policy topics, or types of scholars vary (e.g., Global South versus Global North), so too may the inequities between engagement partners shrink or grow. There may also be several other viable engagement formats beyond the briefings and dinner events we participated in, and these should be explored further.[17] One nimble option for the post–COVID-19 pandemic reality is online platforms. As these formats become more normalized they can bring new voices into the conversation without the risks and costs of travel, including voices that may not speak in our *lingua franca*. Although partnered engagement is not without risks, with sufficient planning, organization, support, and networking the risks can be managed to produce valuable experiences for all.

[16] Unfortunately, members of the ATCC report not having received any follow-up or sustained partnership from the policymakers or organizations that Cristina met with during her stay in Washington, DC.

[17] Even going through the exercise of envisioning partnered engagement may foster greater empathy for partners' interests. It may move academics to reflect on how they represent the views of other actors and what those actors may value in terms of policy engagement.

References

BBC (2014) "Ethics Guide: Duty-Based Ethics." http://www.bbc.co.uk/ethics/introduction/duty_1.shtml.

Berman, Eli, Jacob N. Shapiro, and Joseph Felter (2018) *Small Wars, Big Data: The Information Revolution in Modern Conflict*. Princeton: Princeton University Press.

Bourke, Brian (2014) "Positionality: Reflecting on the Research Process." *The Qualitative Report* 19(33): 1–9.

Center for Civilians in Conflict (CIVIC) (2020) "Having Their Say: Guidelines for Involving Local Civil Society in the Planning, Design, Implementation, and Evaluation of US Security Assistance and Cooperation." October 29, 2020.

Center for Civilians in Conflict (CIVIC) (2021) "Ignoring Iraq's Most Vulnerable: The Plight of Displaced Persons." April 2021.

Chavez, Christina (2008) "Conceptualizing from the Inside: Advantages, Complications, and Demands on Insider Positionality." *The Qualitative Report* 13(3): 474–494.

Chevalier, Jacques M., and Daniel J. Buckles (2013) *Participatory Action Research: Theory and Methods for Engaged Inquiry*. 1st ed. Routledge.

England, Kim V. L. (1994) "Getting Personal: Reflexivity, Positionality, and Feminist Research." *The Professional Geographer* 46(1): 80–89.

George, Alexander L. (1994) "The Two Cultures of Academia and Policy-Making: Bridging the Gap." *Political Psychology* 15(1). Special Issue: Political Psychology and the Work of Alexander L. George: 143–172.

Kant, Immanuel (1781) *Critique of Pure Reason*.

Kaplan, Oliver and Cristina Serna (2018) "Local Strategies in Atrocity Prevention: Self-Protection Strategies are Key to Atrocity Prevention," in *Atrocity Prevention and Peace Building: Key Insights and Lessons from a Global Consultation Convened by Peace Direct*, April 5, 2018. https://www.peacedirect.org/atrocity-prevention-and-peacebuilding/.

Kaplan, Oliver (2013) "Protecting Civilians in Civil War: The Institution of the ATCC in Colombia." *Journal of Peace Research* 50(3): 351–367.

Kaplan, Oliver (2017) *Resisting War: How Communities Protect Themselves*. Cambridge: Cambridge University Press.

Kaplan, Oliver (2020) "Taking it to the Streets: Engaged Research in Political Science." University of Denver working paper.

Mahony, Liam and Luis Enrique Eguren (1997) *Unarmed Bodyguards: International Accompaniment for the Protection of Human Rights*. West Hartford, CT: Kumarian Press.

Merriam, Sharan B., Juanita Johnson-Bailey, Ming-Yeh Lee, Youngwha Kee, Gabo Ntseane, and Mazanah Muhamad (2001) "Power and Positionality: Negotiating Insider/Outsider Status within and across Cultures." *International Journal of Lifelong Education* 20(5): 405–416.

Nye, Jr., Joseph S. (2008) "Bridging the Gap between Theory and Policy." *Political Psychology* 29(4), The Enduring Legacy of Alexander L. George: A Symposium: 593–603.

Peace Direct (2019) "Peace Direct M&E Final Report on Peacebuilder Visits to the US."

Prabhakar, Arati (2023) "Let's Change what's Possible." *Science* 379(6637; March 16): 1069.

Rose, Gillian (1997) "Situating Knowledges: Positionality, Reflexivities and Other Tactics." *Progress in Human Geography* 21(3): 305–320.

6

Should Economists Deceive? Prosocial Lying, Paternalism, and the "Ben Bernanke Problem"

George F. DeMartino

> *Deceit and violence—these are the two forms of deliberate assault on human beings.*
>
> **– Sissela Bok, 1999**

> *The purpose of studying economics is not to acquire a set of ready-made answers to economic questions, but to learn how to avoid being deceived by economists.*
>
> **– Joan Robinson, 1978**

Introduction

Many economists take it for granted that economists lie, and that they should.[1] Many would certainly concur with Anatole Kaletsky (2014) when he asserts, "For just as James Bond has a 'License to Kill' in the Ian Fleming books, so central bankers possess a 'License to Lie'—or, putting it more diplomatically and politely, to make promises about the future that cannot be honored and often turn out to be false." In the view of former World Bank economist Liaquat Ahamed (2009: 363), "What [government economists] have to say about the economy affects its outcome ... As a consequence, they have little choice but to restrict themselves to making fatuously positive statements which should never be taken seriously as forecasts." These observers

[1] Joe Downes, Holden Fitzgerald, and Quentin Good provided wonderful research assistance for this project. Thanks also to Rachel Epstein, Ilene Grabel, Cullen Hendrix, Oliver Kaplan, and Bob McMaster for comments on previous drafts.

George F. DeMartino, *Should Economists Deceive? Prosocial Lying, Paternalism, and the "Ben Bernanke Problem"*.
In: *Speaking Science to Power.* Edited by: Rachel A. Epstein and Oliver Kaplan, Oxford University Press.
 DOI: 10.1093/oso/9780198875185.003.0006

have their finger on what I'll refer to as the *Ben Bernanke Problem*—the problem that arises when truthful speech might be expected to induce damaging outcomes, while lying might be expected to bring about good outcomes. What are we to make of it?

Some noneconomists who examine deception are readily prepared to excuse the practice in economics. It is notable that philosopher Stuart Hampshire, in his discussion of the conflict among ethical principles that often emerges when professionals occupy multiple institutional roles, chooses economics as an example of a profession that can be forgiven for deception. The "Chancellor of the Exchequer [UK Treasurer]," he asserts without reservation, "is not required to respond honestly to questions about a future devaluation of the currency" (cited in Wueste 1994: 3). The implication is that the Chancellor can respond *dishonestly*—no wonder Robinson's (1978) above-quoted epigraph observes that a benefit of economics training is to avoid being deceived. Philosopher Sissela Bok, one of the leading authorities on the ethics of deception, dissents. Discussing the very same scenario Bok (1999: 177–178) acknowledges that while it may be imprudent for government spokespersons to disclose an intended devaluation, the better course is "to refuse comment than to lie." The view of Alan Freeman (2016: 651), an outspoken critic of economists' deceptions, aligns with that of Bok:

> [The] precept "first tell no untruth" is the ethical equivalent, for an economist, of the Hippocratic injunction "first do no harm." If economists unjustifiably claim truth for their statements, and if people act on these statements when different actions could have resulted in better outcomes, then harm has been done, to which the statements have contributed materially.

In fact, there is a tension in how the profession regards lying.[2] On the one hand, it is widely accepted in the profession that there are occasions when economists deceive and even outright lie—situations when they believe doing so is warranted by the public interest. On the other hand, there are limits to the nature of lies that the profession is willing to accept (see Mayer 2016). But where are they? J. M. Keynes is reported to have stated to Abba Lerner, perhaps in jest, that "It's the art of statesmanship to tell lies but they

[2] For present purposes it suffices to equate deception and lying, and I use the terms interchangeably. For other purposes it is not. In her treatise on lying, Bok distinguishes between them this way: "I shall define as a lie any intentionally deceptive message which is *stated*." Deception is a "broader category," which also comprises purposely misleading "gesture," "disguise," "means of action or inaction," and even "silence" (Bok 1999: 13). Here I largely restrict my attention to economists' statements, though I define the term widely to include not just their findings but also, for instance, how they report their findings, the claims they make for their data, the disclosures they do and do not make about their conflicts of interest (Flitter et al. 2010; Carrick-Hagenbarth and Epstein 2012), and so forth.

must be 'plausible lies'" (Colander 2019: 70). Surely that cannot be the right criterion to establish the limits to deception? But, lacking a field of professional ethics—a tradition of careful inquiry into ethical conduct—each economist is left to be his or her own judge of which lies are appropriate to tell and which are not. There are no standards to guide economists' behavior, no debate in the economics profession (either among academic or applied economists), and no learning from economists' decisions, ex post.[3]

While preparing to write a review essay on new books by Dani Rodrik and Joseph Stiglitz for *Ethics and International Affairs* (DeMartino 2019), I was struck by the urgency with which Stiglitz and especially Rodrik allege that economists deceive. Both claim that, over many years, academic and applied economists deceived the public in support of the neoliberal agenda. Here are some assertions from just four pages in Rodrik's *Straight Talk on Trade* (2018: ixff; emphasis added). All appear in the Preface—they are not buried deep in the book. Rodrik says that trade economists "*overstated* the magnitude of aggregate gains from trade deals" and "*minimized* distributional concerns"; "*endorsed the propaganda*"; "*parrot[ed] the wonders* of comparative advantage"; demonstrated a "*reluctance to be honest*"; and "*fail[ed] to provide the full picture.*" And that's not all. Trade economists purposely "*failed to engage* distributive justice issues"; "*shade[d] their arguments*"; worked as "*academic boosters*" of market fundamentalism; and should have been "*more upfront*" and presented "*a more honest narrative.*" Rodrik notes the disconnect between the seminar room, where economists acknowledge and engage openly with the limitations to their science, and public forums, where economists oversimplify what their models tell them, exaggerate their capacities, and cherry-pick evidence so as to have maximum influence over public perceptions.

How should one account for the subterfuge? Standing on "the side of angels," says Rodrik (2018: ixff), the free traders were reluctant "to provide ammunition to the barbarians"—those protectionists who would exploit any equivocation among economists about the virtues of trade. Later, Rodrik argues that "in their zeal" to impress their audiences, economists "skip over the real-world complications and nuances" and "downplay the diversity of intellectual frameworks." Stiglitz (2018) adds that the profession told a "little

[3] Though it is not my focus here, my sense is that much of the same is true in political science. Several of the chapters in this volume wrestle, in one way or another, with some of the issues I raise here. For instance, Epstein (Chapter 7 in this volume) explores the case of NATO enlargement (in 1999 and beyond) and finds that various security scholars overstated their claims, overgeneralized their findings, and failed to acknowledge their political priors and the limits to their objectivity. All of these strategies entail deception—of those whom scholars advise and their broader audiences—and, often, self-deception.

white lie"—that trade creates jobs—and continually "oversold" globalization. Elsewhere, Stiglitz (2016) goes much further in challenging the profession's associated ethical failures.

In my reading the allegation is clear: during a period of global economic transition toward neoliberalism—an epochal transformation that altered the life circumstances and prospects of billions of people the world over—*the economics profession failed in one of the most egregious ways that a profession can fail.* The profession *lied* to the public so as to promote the success of a regime that the profession believed to be best. The subterfuge was intended, inter alia, to hide the fact, of which economists were well aware, that the transition would generate substantial harm to large numbers of people. The worry, of course, was that an honest, public engagement with the harms, and with the confounding complexities and uncertainties of the project, would undermine the project's chances of implementation and success.

I take as my principal theme here the following question: Is it appropriate that the economics profession should give itself license to violate a principle of truthfulness that is by now universally recognized as an important dictate across the professions and among those scholars who have probed the issue most carefully?[4] The question can be reframed as follows: Is there a case to be made for ethical exceptionalism for economics? If so, what forms of deception are to be deemed appropriate, and which are to be recognized as inappropriate? Alternatively, is there something ethically worrisome about the warrant economists give themselves to deceive?

In what follows, I lay out the most compelling defense of economic exceptionalism, one that I think accords with how economists tend to view the matter. I then examine the case for economic deception and identify some of the ethical challenges that arise once a profession explicitly or implicitly comes to accept that it is appropriate to lie. In particular, I emphasize the difficulties associated with setting the boundaries between permissible and impermissible lying within a profession that tolerates deception. I also emphasize the fact that deception corrodes the trust—within the profession, and between the profession and others—upon which the profession's efficacy depends.

[4] In Bok's view (1999: 19–20), lying undermines agency by conveying the false idea that there are no alternative courses of action available to an agent other than the one favored by the deceiver, and can distort agents' sense of the depth of uncertainty concerning alternative courses of action. Galtung (1969: 169) examines "violence that works on the soul; where the latter would include lies, brainwashing, indoctrination of various kinds, threats, etc. that serve to decrease mental potentialities." Grant (2012: 47) argues that deception should be understood as a kind of force or coercion rather than persuasion.

Prosocial Lying: The "Dirty Hands" Defense

The best defense of professional deception in the public domain involves what Levine and Schweitzer (2014, 2015) call "prosocial" lying: lying in service not of the deceiver, but of the deceived.[5] Others refer to it as paternalistic lying, for reasons examined below. The defense of prosocial lying recognizes truth-telling as a *prima facie* obligation that is nonetheless sometimes overcome by various exigencies. Liars "dirty their hands" to promote social good.[6] The "dirty hands" defense often entails recognition of a conflict between deontological and consequentialist ethics, where a moral imperative is violated to bring about the best overall outcome, such as maximum social welfare.[7] In this view a lie that brings about better aggregate outcomes than does truth-telling is not just permissible but is in fact morally required. Professionals who let their ethical squeamishness prevent them from lying in service of the public good are, on this account, ethically deficient and even cowardly.

Prosocial lying could be considered appropriate in a range of circumstances and under a range of conditions. Here I list ten circumstances that might justify deception. The list is intended to be illustrative, not comprehensive (see Table 6.1).[8] First, prosocial lying may be taken to be essential

Table 6.1 Justifications for prosocial lying in economics

Crisis conditions
Immediate risks, fleeting opportunities
Epistemic asymmetry between professionals and laypersons
Seductive but harmful alternatives
Faulty modes of communication and instruction
Defensive, deception-countering lies
Adversarial ethics
Corrupt or deficient decision-makers and decision-making processes
Research dictates
Consent of the deceived

[5] "Prosocial lies involve the transmission of information that misleads and benefits a target" (Levine and Schweitzer 2015: 89).

[6] On the "dirty hands" problem, so named after the play by Sartre (1960), see Walzer (1973), Thompson (1987); and Winston (1994). For "dirty hands" in the context of deception, see Bok (1999).

[7] In contrast, Winston (1994) holds that the "dirty hands" problem is better theorized in terms of conflicting duties, where an agent's actions to meet the duties to some persons violate his/her duties to others.

[8] For instance, Olsen (this volume) provides insight into a kind of justification not included in this list: the urgency to preserve working partnerships with those organizations and actors that scholars seek to serve. In that context, bad news from the scholar in the form of findings that challenge the underlying assumptions that drive policymaking might be unwelcome, and might threaten the partnership. Is there

under crisis conditions where decisions must be made and behaviors adjusted immediately, when there is insufficient time for lengthy dialogue between the professional and laypersons, and when the consequences of failing to take certain actions are dire.[9] In such cases, the argument runs, the professional must tell the lies necessary to avert disaster—leaving it until later (if ever), after the crisis has passed, to defend the crisis-induced lie on prosocial grounds.

A second, and related, justification is also grounded in temporal urgency given by immediate risks of harm or fleeting opportunities for benefit. Mill's (1859/2001: 88) discussion of the right of a bystander to use temporary force to restrain a person from walking out onto an unsafe bridge provides a framework for illuminating the matter. It can be argued that one should employ deception in the moment to stop an individual from harming themself, if no other (less damaging) means are available to stop that person. But the deception may be maintained only long enough to inform the individual of the present risk. Alternatively, a fleeting opportunity for social benefit may be lost without the abrupt insertion of deception into the flow of events. Like the crisis justification, deception is used here to freeze action or to alter behavior in the moment, while at the same time permitting a full, *honest* discussion among the parties of the situation after the fact. The use of deception to permit the circumstances for honest exchange is strange on its face—but not inconceivable nor, perhaps, uncommon.

A third justification foregrounds the enduring epistemic asymmetry that characterizes the professional–layperson relationship (DeMartino 2013, 2022). If the epistemic gap between economists and noneconomists is believed to be unbridgeable, then it might be argued that the professional is absolved of the duty to speak truthfully since doing so would fail to communicate sensibly the salient features of the situation or the consequences of the decisions to be taken. It is important to note that, unlike the crisis or temporal constraint justification, this one licenses ongoing deception so long as the epistemic asymmetry persists. An imperfect analogy involves the parent–child relationship, where parents lie to induce appropriate behavior in children when truthful discussion would be pointless and would fail to bring about the right result.

Fourth, a choice environment might include one uniquely best but apparently unattractive option, and a seductive but harmful alternative. Decision-makers might be drawn to the attractive alternative, unaware of

a case, then, for deception to preserve relationships that are essential to the scholar's work, and even to social betterment?

[9] See the discussion in Bok (1999, ch. 8) of the crisis justification for deception, and the dangers thereof.

and perhaps unable to fathom its severe downside. In such circumstances the economist may steer decision-makers to the less attractive option by simply not presenting the seductive option. The greater the attractiveness of the alternative, the stronger is the case for "there is no alternative" (TINA) deception—deception that constrains the choice set to include just the option deemed appropriate by the economist. Since it is difficult in the policy context to claim that there really is just one available option, TINA deception is typically associated with ersatz cost–benefit analysis that exaggerates the benefits of the preferred option and the costs of any alternative.

Fifth, it may be argued that prosocial lying is warranted when the existing modes of communication and instruction do not permit effective transmission of knowledge. Here, the issue is not that the epistemic gap is unbridgeable. Rather, the conditions that would permit the bridging do not obtain, for whatever reason. Ideally, the situation should be reformed to permit effective professional–layperson communication and education; but, in the meantime, professionals are justified in telling prosocial lies to bring about the situation that they would otherwise be able to bring about through truth-telling.

The sixth entails deception-countering lies. It involves lying to defeat the lies others tell. Under ideal discursive arrangements the discussion of existing circumstances, strategy options, distributional consequences, and the like would be conducted by participants operating in good faith who avoid deception. But, in fact, agents often tell self-interested lies, lie to themselves (especially when their interests are at stake), or engage in prosocial lying for any of the reasons given here. In such cases, prosocial actors may find that they can only prevail in disputes with their dishonest adversaries by telling more compelling lies.

Seventh, a variant of the claim that economists can lie to defeat others' lies adopts the idea that the public arena is contentious, and that each party in public policy mêlées has a right to present the case that best supports its position, even if that means cherry-picking or shading the evidence in its favor. Here, the economist adopts the adversarial ethic of the legal profession (DeMartino 2011). The adversarial economist presumes that the best decisions will emerge from the battle of opposing viewpoints. The economist is therefore to choose just those facts that sustain the position of its client, leaving it to the other side to present the opposing case (and data) as best it can. Here is how an economist who works only for industry put it in an interview I conducted when doing research for *The Economist's Oath* (2011):

> While an academic economist places all the evidence he can find on both sides of the scale and sees how things balance out, the economic consultant's job is to load up as much evidence as he can find on just one side of the scale—that side that advances the client's interests. But all this evidence must nevertheless be legitimate and honest.

Eighth, prosocial lying might be defended on grounds that decision-makers whom economists advise are corrupt, have deficient values, or hold mistaken beliefs about those they represent that obstruct prosocial decision-making. Alternatively, they are enmeshed in decision-making structures that prevent prosocial outcomes.[10] In such cases the expert should intervene strategically, this argument goes, including telling prosocial lies, in order to induce the morally impaired decision-makers or the technologically impaired decision-making mechanisms to produce the right results. For instance, an economist serving as a consultant might attempt to persuade narrowly self-interested decision-makers that the policy choice that the economist knows to be best for society will also benefit the decision-makers.

Ninth, a profession may come to think that deception is necessary in research that engages human subjects. This issue arises, for instance, in behavioral economics, where economists run experiments in labs. Here, an economist might find it convenient to use confederates who present themselves as research subjects but who in fact covertly play prescribed roles to elicit the reactions of actual research subjects. Or, an economist might want to know how the behaviors of research subjects are altered by their perception that they are one of a very large number of subjects when in fact they are not, so as to ascertain tendencies toward free riding (Wilson 2016).

Finally, it may be argued that those to whom experts lie have consented to be deceived, or that they would invariably consent to be deceived, were they to be in position to do so. A patient may ask his physician not to tell the truth about the likely success of a dangerous treatment regimen, when in fact there are no viable treatment options. Here, the physician might be licensed to tell benevolent lies—and, in fact, doing otherwise might be deemed callous. But economists do not treat clients one-on-one in this way. Economists' actions (when they affect anyone at all) affect many thousands and even millions all at once. In that context, the consent to be deceived cannot be broadly obtained

[10] Sisk (Chapter 4, this volume) describes a type of case—peace building in deeply divided societies—where this kind of justification for lying could be offered.

and can only be presumed. Those who proffer this defense of deception must be prepared to demonstrate that the evidence in favor of that presumption is very strong.

For these reasons, the case for prosocial lying appears at least plausible, and perhaps even compelling. And yet it is not widely embraced across professions. What is to be learned, then, from the skepticism of other professions about lying for the good of others?

The Case of Medical Practice: The Paternalistic Hippocratic Tradition

Medical practice in the United States during the nineteenth and twentieth centuries was driven by the Hippocratic principle of harm avoidance. The Hippocratic doctrine is fundamentally paternalistic in spirit. It is characterized by layperson deference to physicians' authority, and the substantial latitude of the physician regarding truth-telling. An early version of the Hippocratic Oath included the following passage:

> I will follow that system of regimen which, *according to my ability and judgement*, I consider for the benefit of my patients, and abstain from whatever is deleterious and mischievous. (emphasis added; DeMartino 2011)

Nowhere does the Oath (nor the broad Hippocratic tradition; see Shuster 1998) recognize the autonomy or integrity of the patient nor expressly prohibit lying. Neither does "the prayer of Maimonides, Percival's Ethics, or the early codes of the [American Medical Association] or the World Medical Association (Veatch 1984) ... Outside agents—including the patients themselves—were simply not deemed qualified to participate in the formulation of ethical behavior within the profession" (Wolpe 1998: 39). In Hippocrates' own words,

> The art of medicine has three factors, the disease, the patient and the physician. The physician is the servant of the Art. *The patient must cooperate with the physician in combating the disease.* (cited in Shuster 1998: 974; emphasis is Shuster's)

Under the Hippocratic tradition it is permitted and perhaps even required for physicians to deceive their patients when they think deception will minimize pain and suffering—or when, in the physician's view, lying is the best way to induce patients to change their behavior or agree to a course of treatment that the physician prefers.

The Patients' Rights Movement

The patients' rights movements of the 1960s and 1970s targeted Hippocratic ethics. In the US context the courts intervened decisively, altering the nature of medical practice and ethics. The landmark decision is *Canterbury v. Spence*, 1972. Here the DC Circuit of the US Court of Appeals overturned the physician's authority to deceive, finding that "'the patient's right of self-decision shapes the boundaries of the duty to reveal'" (Sharpe and Faden 1998: 54). A decade later (1974) the National Research Act established the Institutional Review Board (IRB) procedures that protect the rights of research subjects, which continue to the present. By the 1970s, Sharpe and Faden (1998: 67) report, "professional paternalism was increasingly challenged by the publicly and politically-forged ethos of patient self-determination."

Today, professional deception in medicine is taken to be deeply problematic. Among other things, it is recognized as a form of coercion. Agents coerce others not just or even principally with a gun to the head, but also by shaping the beliefs of others about available courses of action, and the viability and desirability of the available options. When professionals induce others to do what they want them to do by lying to them "for their own good," they deprive them of their personal autonomy. Lying therefore runs afoul of the Kantian imperative to treat others always as an end, and not simply as a means.[11] This form of coercion is far more pernicious than the gun to the head since deception deprives the coerced agent of the opportunity to know that they are, in fact, being coerced. *They are doubly wronged.* A deceived individual suffers a loss of autonomy *and* the opportunity to resist, or to learn. This bears emphasis: the ethical grounding for prosocial lying is essentially paternalistic. *It presumes that the professional knows best and is licensed to do what it takes to promote social betterment, even when doing so requires coercion of those whose betterment is targeted for enhancement.*

[11] Dworkin (2005) writes: "To deny an adult the right to make their own decisions, however mistaken from some standpoint they are, is to treat them as simply means to their own good, rather than as ends in themselves." A fuller examination than I provide here would have to engage the possibility that one could ground a strong defense of prosocial lying in virtue-ethics—perhaps strong enough to overcome the Kantian objection. I thank Robert McMaster for pressing this point on me. This issue also arises in the model of partnered engagement (see Kaplan, Chapter 5, this volume).

One indicator of the ethical shift in medical ethics is the altered content of contemporary medical oaths.[12] Consider this passage from the Yale Physician's Oath, drafted during the 1990s by medical students and the university Chaplain:

> I will respect the moral right of patients to participate fully in the medical decisions that affect them. I will assist my patients to make choices that coincide with their own values and beliefs. (cited in DeMartino 2011)

As the excerpt suggests, the physician's authority to undertake a treatment regimen is now substantially constrained by the need to engage the patient as a partner in the treatment plan. This oath places paternalistic strategies in ethical doubt: consent secured on the basis of deception can hardly be considered informed. Under the dictate to respect the integrity and autonomy of patients, such consent is therefore deemed invalid. The oath codifies a presumption that the *patient*, and not the physician, should decide how much and what kind of information they will receive, for instance, and the demand to be fully informed must be honored no matter how damaging that information may be to the patient's physical or psychological wellbeing.

Other professions have followed medical practice in moving away from prosocial lying and toward the dictate of truth-telling so as to respect the integrity and autonomy of laypersons. The economics profession lags in this respect, remaining anchored to the paternalistic ethos. Returning to the question posed earlier: Is this as it should be? Can a compelling case be made for economic exceptionalism? Is economics unlike other fields in ways that make paternalistic coercion, through lying, ethically appropriate?

Economic Exceptionalism and the "Ben Bernanke Problem"

Let's begin with a case of prosocial lying that, I think, many (if not most) economists (and many non-economists) would accept as legitimate. I refer to the utterances of Ben Bernanke in the period leading up to and during the financial crisis of 2008. Bernanke is a professor specializing in the Great Depression who was appointed by President George W. Bush as Chairman of the Federal Reserve in 2006 and was awarded the Nobel Prize in Economic Sciences for his work in 2022 (Figure 6.1). Already upon his appointment

[12] This discussion draws on DeMartino 2011.

Figure 6.1 Swearing-in ceremony for Dr. Ben Bernanke as the Chairman of the Council of Economic Advisors, 2005.
Courtesy Archival Operations Division, NARA

signs of deepening financial fragility tied in large part to the US housing market were emerging. I take this to be the paradigmatic case that involves many of the defenses of prosocial lying reviewed earlier. I should emphasize, however, that the case for lying is not unique to central bankers, and central bankers are not in fact my principal interest in connection with lying. The same arguments that sustain central-banker lying apply equally to the practice of economists in every sector where they work, and from the local to the international spheres. This represents a problem for the profession, and might undermine the case even for central banker deception.

Counterintuitively, Bernanke's statements to Congress and the public became more reassuring as economic conditions deteriorated. In the early months of 2008, we find assertions of increasing confidence as the financial collapse nears. On May 15, 2008, in comments to the 44th Annual Conference on Bank Structure and Competition, Bernanke spoke candidly about financial risks:

> The financial and credit market turmoil that began last summer has raised a number of significant issues of public policy, including questions concerning the maintenance of financial stability, the supervision and regulation of financial

> institutions, and the protection of consumers in their financial dealings ... The problems with subprime mortgage underwriting were disguised for a time by the continued appreciation in home values. As long as house prices kept rising, subprime borrowers saw their home equity increase and were often able to refinance into more-sustainable mortgages. But when house prices began to stagnate and then fall, many subprime borrowers found themselves trapped in mortgages they could not afford. (Bernanke, May 15, 2008)

Just two weeks later Bernanke presented a more sanguine account. On June 3, even though economic conditions had worsened, he reported to the International Monetary Conference:

> The current economic and financial situation reflects, in significant part, the unwinding of two of these longer-term developments—the housing boom and the credit boom—and the continuation of the pressure of global demand on commodity prices ...
>
> The Federal Reserve is pursuing its objectives through several means. First, we have eased monetary policy substantially and proactively to address the sharp deterioration in financial conditions and to forestall some of the potential adverse effects on the broader economy. Our decisive policy actions were premised on the view that a more gradual reduction in short-term rates could well have failed to contain the financial and economic problems confronting us. *For now, policy seems well positioned to promote moderate growth and price stability over time* [emphasis added].

The same week (June 9) he also discussed financial uncertainty as if it represented nothing more troublesome than a minor impediment to growth:

> I would like to provide a brief update on the outlook for the economy and policy, beginning with the prospects for growth ... [Although] activity during the current quarter is likely to be weak, the risk that the economy has entered a substantial downturn appears to have diminished over the past month or so. (cited in Ip and Reddy, June 10, 2008)

Then, in July 2008, presenting the central bank's semiannual report on monetary and economic policy to the House Financial Services Committee, he asserted that "the GSEs [Government-Sponsored Enterprises] are adequately capitalized. They are in no danger of failing" (Herszenhorn and Weisman, July 17, 2008). Just two weeks later, Warren Buffet signaled his concerns about the GSEs; a month later (September 7th), the US government was forced to bail out two principal GSEs: Fannie Mae and Freddie Mac. The Treasury was authorized to purchase up to $100 billion in preferred stock and mortgage-backed securities to save the institutions and prevent financial panic.

I presume that by mid-2008 Bernanke faced extraordinary pressure to calm investors and prevent financial collapse. I submit that during the period leading up to the crisis his speech was increasingly strategic and performative rather than descriptive. By then, there were ample signs of financial fragility. Bernanke's interventions were directed toward sustaining confidence so as to prevent a crisis. His utterances at this point were not truthful; they were deceptive. If that presumption is correct, then his behavior reflected prosocial lying.

It might seem obvious that in this situation Bernanke was right to deceive—his purpose was to prevent economic crisis, and one chief means available to him to achieve that goal was deception. Here and in other similar situations where economists lie, deception takes the form of "strategic speaking." In the context of central banking, the fairly recent shift from fewer to more public pronouncements and the associated shift to what is called "forward guidance" open the door to increased prosocial lying, even in non-crisis situations. Central bankers increasingly speak privately and publicly with lawmakers, financial investors, and other influential economic actors, and to the public. In this context, "open mouth" interventions represent strategic speech intended to produce a particular result. Speeches are one of the means by which central bankers look to exert effects on the economy. A growing research field today is exploring whether, how, and to what degree forward guidance affects economic agents' behavior and helps to achieve the purposes for which it is undertaken, and with what risks (e.g., Kool and Thornton 2012; Kliesen et al. 2019; Haldane et al. 2020).

Strategic speaking often takes the form of conveying information, but its purpose is not primarily to inform. It is to bring about a variety of concrete effects. The test of the speech is not the truthfulness of its content, but rather its contribution to the effects it targets. The goal is to alter beliefs in order to affect behaviors and, in that way, to generate particular aggregate outcomes. Central bankers may characterize the economy as healthy, for instance, not primarily to describe what they see or what they believe to be true, but as a strategy to bring about stability.

Strategic speech is ethically complex. The *truth* of the strategic utterances of central bankers depends, in significant part, on the *ultimate effects* of the statements, which they cannot know when they make them. What makes the statement true or false ex post depends on whether the statement has its intended effects, and that in turn depends in part on a wide range of factors that the banker cannot know at the time of the statement. It is in this sense a calculated falsehood that may yet prove to be validated by subsequent events.

Consider a generic statement of the sort "Financial markets are stable." The statement should be interpreted not as descriptively true, but as a request to market actors—that is, "Please act on the belief that markets are stable." But why does the central banker then not just make that request, which would dispense with dishonesty altogether? The banker must believe that the request alone would not suffice to induce the hoped-for behavior by market actors. In fact, the request might even have the opposite effect by alerting market actors to the central bank's perception of the threat of instability. Only a declarative statement of the form "financial markets are stable" may be thought to have a chance of achieving the goal of stabilizing markets. The foregoing implies that the performative force of a strategic statement is what matters, not its truth content. But notice: the abandonment of truthfulness as a criterion for assessing professional utterances opens the door to more significant deceptions. When a bigger lie or a longer chain of lies is needed to get the job done, greater deception would appear to be warranted. Is that something the profession should tolerate?

Assessing Economic Exceptionalism

Prosocial lying in economics is clearly paternalistic, and coercive since its intent is to alter perceptions and behaviors not through reasoned discussion with others about risks and strategies for meeting them, but by deception. Lying in economics therefore violates the autonomy of laypersons populating the economy.[13] It is an instance of the "dirty hands" justification cited earlier, whereby an agent feels compelled to violate an important value so as to bring about what the actor deems to be a more important social good (Walzer 1973; Thompson 1987; Winston 1994). So, the defense of lying must demonstrate that the goal to be achieved justifies the ethical transgression: the diminished autonomy of laypersons populating the economy. The defense must also sustain the claim that it is rightfully within the economics profession's authority and competence to decide when it is and when it is not appropriate to make

[13] This is true unless one presumes that agents have granted to economists the authority to lie to them when economists believe it is in their best interests to do so, as per the earlier discussion of justifications for prosocial lying. A problem is that the exercise of securing explicit permission to lie would undermine the persuasive force of the lies economists tell. Hence, central bank prosocial lying, if it is to be effective, must be obscured by the trappings of truth-telling. See Bok (1999, ch. 14) on the deep problems associated with implied consent for deception.

the decision to coerce. This is precisely the position of the medical profession prior to the *Canterbury* decision in 1972: it took it upon its own shoulders to decide when it was authorized to lie.[14]

Who Can Lie, and in Which Ways?

If economists accept deception by central bankers as appropriate to promote the social good, the profession must then answer a range of difficult questions. First, are economists more broadly licensed to deceive regarding their findings, certainty, or expertise in order to promote what they believe are good economic outcomes? It would seem that they are. Economists operate in many arenas where it can be argued that deference to their judgments will promote social wellbeing. There do not seem to be compelling grounds for granting central bankers license to engage in prosocial lying while at the same time denying that license to other economists. But, then, is there no line to be drawn among economists and situations where prosocial lying is and is not appropriate? Take, for instance, the economics professor who knows that society would be better off were all economists, including those in the seats of advanced graduate courses, to believe that only the predominant approach to economics is correct. Is the professor warranted in lying about the value of other approaches, dismissing them out of hand, so as to eliminate them from serious consideration and to ensure that students not become confused by the contending claims of diverse perspectives? George Stigler thought so, as Colander and Freedman (2019: 210) recount: "One of the rationales provided by Stigler for dropping History of Thought from graduate education was the danger of providing young minds with too many plausible alternatives. He required missionary soldiers, not ranks of puzzled philosophers."

A separate, though related, question concerns the *forms* of permissible lies. In the case at hand I have discussed situations where central bankers' deceptions take the form, in part, of an exaggeration of their knowledge and expertise, as well as an overestimation of the capacities of the institution a central banker leads, so as to induce deference by others to bring about good social outcomes. As Bok (1999; 20) warns us, deception regarding the level of uncertainty "is one of the main ways to gain power over the choices of those deceived." I would add that it is perhaps the most common form of deception in economics, and the most tolerated.

[14] The problem of cognitive bias is acute in this context. See Bok (1999: 26).

But if economists can deceive in *that* way, is it acceptable for them to deceive in *other* ways to bring about the very same social objectives? What about a case in which an economic consultant exaggerates both the benefits of a preferred policy option and the deficiencies of the alternative proposals that the consultant opposes, by manipulating a cost–benefit analysis? Is *that* form of prosocial lying permissible if the economist believes it is the only means to ensure good economic decision-making? What if that consultant purposely excludes policy options so as to prevent the receiver of the advice from choosing poorly?

Deception by academic, public, and private sector economists can also take the form of falsifying reports or data, or misleading audiences concerning the dependability of the findings one reports. Economists may engage in extensive p-hacking, or rigging their analyses to show significant results, but then fail to disclose the practice (Mayer 2016). Experimental economists may similarly treat trials that contradict proposed hypotheses as "pilots" and record as actual trials only those that confirm the initial hypotheses (Roth 1994). Are government economists in statistical agencies warranted in misreporting data concerning consumer confidence, job creation, inflation, unemployment, trade balances, and other indicators of economic flows and outcomes, if they can get away with it, and if they have good reason to believe that doing so would promote some important social objective? For instance, China is widely regarded as having routinely fabricated economic data so as to sustain foreign direct investment, popular support, and political stability—all of which may have been necessary for the success of its economic model and the dramatic reduction in poverty rates it has achieved over the past several decades. What if economists believe that those forms of deception are the best way to bring about desirable economic outcomes? Are those forms of deception also then to be deemed legitimate?

There is no doubt that economists would be disturbed by the prospect of central bankers and government economists fabricating data, even if it is done to promote social betterment, just as they would be disturbed to learn that an academic economist has done so. But the question we must then confront is this: *Why are these kinds of lies illicit, when others are legitimate?* Where do we draw the line, if in fact there is a line to be drawn? Within the consequentialist ethics of standard welfare economics there is little to no ethical difference between these kinds of lies and the other lies considered previously, provided they all achieve prosocial purposes. But, then, on what grounds might economists approve of one kind of lie yet reject others?

What is at issue for economists making these judgments is not an ethical principle at all but rather, I conjecture, professional pride. When Bernanke lies about his grasp of the unknown and unknowable, economists are in on the deception. *Economists see "the wink."* To the degree that the deception succeeds, the profession as a whole is aggrandized by the lie. The profession is validated and empowered, and feels justified *in his deceiving noneconomists.* But economists, I also submit, are intolerant of lies when *they themselves* are among those being duped. The distinction is this: *Economists demand that Bernanke recognize the agency of economists, even as they're prepared to sacrifice the agency of others.* Bok (1999: 23) gets at the double standard in terms economists will easily understand: "They would prefer, in other words, a 'free-rider' status, giving them the benefits of lying without the risks of being lied to. Some think of this free-rider status as for them alone. Others extend it to their friends, social group, or *profession*" (emphasis added).

The foregoing discussion of the purposes, defense, and complexities of prosocial lying leads to an ethical fork in the road. One can argue that the case for economic exceptionalism fails, but then be prepared to confront the possibility of diminished social wellbeing that might result from truthfulness. That is the fork taken by the medical profession and other professions, like engineering, journalism, and the law. Each of those professions looks to prevent deception in most professional conduct. Alternatively, if one holds that the case for economics exceptionalism succeeds, then one faces another fork. One must either be prepared to accept that *all* prosocial lying is permissible in the profession—by all economists, and in all forms, when they have good reason to think that lying will generate better effects than truth-telling—or to draw an ethical line between permitted and prohibited forms and instances of prosocial lying.

It is tempting to think, for instance, that the profession can at once exploit prosocial lying while managing any collateral damage by reserving the right to deceive for high-profile economists such as central bankers whose utterances have the widest impact, and perhaps only in moments of looming crisis. But that standard is uncompelling. The work of economists operating at the municipal level can affect the lives of many millions (more than the populations of small countries). Is an economic consultant for the city of Denver (with a metro-area population of 3 million) to be licensed to engage in prosocial lying, but a central banker in Iceland (population 365,000) to be required to speak truthfully because of the relative size of the populations they serve? It is difficult to specify the grounds for judging their conduct differently. Or is the line to be drawn between lies about one's expertise, confidence, and capacities on the one hand, and fabrication of one's data and lies about one's findings on the other? Is the line to be drawn perhaps between those lies that do and do not deceive other economists? I am aware of no viable ethical standards that support the drawing of any of those lines.

Prosocial Lying: Ersatz Economics and the Problem of Trust

In researching the book *The Economist's Oath* I interviewed many economists who worked outside of academia. I asked about the challenges they face in trying to do good work, which I defined for them simply as work with integrity—work they themselves believed in.

One theme that emerged with particular frequency is that economists are routinely pressed into doing what I would call "ersatz economics": work that they know is deeply deficient and even false; work that they themselves do not believe in. For instance, economists are sometimes expected to produce research and reports ex post to sustain policy decisions that are already taken (see also Olsen, Chapter 3, this volume). This is equally true in the public and private sectors. Here is one prominent example, reported in a paper by Allen (1977), which is consistent with but more dramatic than what I learned from my respondents. A new member of the Council of Economic Advisors, which advises the US President, was instructed to produce (in *six hours*) an estimate of "the balance of payments impact of a West Coast dock strike" for federal government attorneys who would "seek a Taft-Hartley injunction in court the next morning." He continues: "The legally specified basis for the injunction was demonstration of [the] existence of a national emergency. So, first the Administration made the decision to claim a national emergency, in order to obtain the injunction, and then economists were to provide supportive data and analysis" (Allen 1977: 52).

Is that sort of deception acceptable if the complicit economists believe that preventing the strike would promote social betterment? In this case the economist was pushed to produce fake economic analysis. This is not terribly dissimilar to other forms of deception examined here. It would seem that ersatz economics could be easily justified on the same consequentialist grounds that economists are apt to use to judge other forms of professional deception.

But ersatz economics is particularly difficult to defend even on consequentialist grounds, provided one takes an expansive view of the consequences. A profession that is willing to do and accept ersatz science is not apt to retain the trust on which its authority depends.[15] When government economists, economic consultants with public and private sector clients, forensic economists, economists attached to political campaigns or ostensibly nonpartisan think tanks, or even academic economists produce work without integrity—work that is intended to deceive, work that those generating it themselves do not believe but take it to be good for others to believe—then the profession cannot expect to retain the trust of those its lies are intended to serve (Figure 6.2).

In my interviews I found that economists at all these levels are often in positions where they are pressured or even inclined to generate ersatz economics, often for the public good (e.g., see also Caballero Reinoso, Chapter 8, this

Figure 6.2 Professional and institutional trust is fragile, and can be fractured by prosocial lying.
Lane V. Erickson/Shutterstock

[15] Does deceit always damage trust? The evidence is mixed. Levine and Schweitzer (2015) find, through experimental research, that prosocial deception can in fact enhance trust provided those being deceived are aware of the good intentions of the deceivers. The findings provide little insight for situations involving economic practice, where those affected by economists' behavior do not have the close interaction necessary to infer motivations. "In practice," they write, "both motives and the link between acts and outcomes may be difficult to gauge. In some cases, people may even attribute selfish motives to prosocial acts." They also recognize that "it is also possible that prosocial lies have detrimental long-term consequences" (103). In contrast, Lupoli et al (2018: 31) find that "well-intended paternalistic lies may backfire," with research subjects punishing even those deceivers who tell paternalistic lies that contribute to their welfare.

volume). Several had stories to tell of their own compromises with the truth, and of the behavior of other economists. One effect is that economists do not always, or even generally, trust the findings of other economists. We see this in the allegations by Rodrik and Stiglitz discussed at the outset of this paper; by Freeman (2016), Leamer (1983), Mayer (2016); and by the many other economists who have called out their colleagues for deceptive practices. In my interviews, respondents warned me never to trust a development industry economist; never to trust a government economist; and, as one former member of the Council of Economic Advisors, told me, "Never trust an economic consultant." He immediately noted that earlier in his career he himself had been an economic consultant. A DC think-tank economist warned me that those in her line of work were forced to "speak more loudly" on the issues than they otherwise would—exaggerating their findings and their confidence in them—so as to sustain the support of their funders, and to respond to the inflated claims of think tanks at the other end of the political spectrum. Candid conversations like these compel the conclusion that deception of various sorts is routine rather than exceptional in the profession.

There is a final, most difficult hurdle that the defense of economic exceptionalism must overcome if it is to be sustained. A profession that tolerates deception—prosocial or otherwise—confronts the risk of the loss of trust. A profession that accepts prosocial lying risks its own delegitimization, even in the eyes of its practitioners. One might conclude, simply, that it deserves the distrust and that it alone suffers the consequences. But trust is an essential contributor to a profession's ability to promote social wellbeing. A profession that loses trust loses efficacy (Hardwig 1994). Moreover, it can also contribute to a popular narrative that is more broadly hostile to expertise.

Conclusion

It might be expected that an examination of prosocial lying in economics would generate a clear bright line between legitimate and illegitimate deception. I do not offer that bright line. The goal has been more modest. Having articulated what is, essentially, a paternalistic case for prosocial lying, I have attempted to reveal its downsides. I conclude by highlighting four chief concerns. The first is that when economists lie for the presumed good of others, they presume they can know just what all those others value, or should value. But if in the clinical setting it is now widely understood that even a family physician who knows the patient intimately cannot know sufficiently well the patient's values and goals, it is entirely implausible for economists to know the values and goals of the many thousands (or millions) that their work affects. Second, when economists induce behaviors through deception, they undermine the autonomy of those they coerce. Deception of this sort short-circuits learning or other reforms that might allow for more honest expert–layperson communication in the future. Third, having tolerated deception of various sorts, the economics profession has forfeited trust among its members. Economists too often come to accept findings that they find attractive, and discount the rest—alleging illicit motives or incompetence of those who offer unwanted findings.[16] Finally, a tolerance for lying, prosocial or not, undermines trust in the profession among noneconomists.

Bok takes what might be thought of as a general equilibrium analysis to the problem of lying. It should therefore resonate with economists:

> [T]he most serious miscalculation people make when weighing lies is to evaluate the costs and benefits of a particular lie in an isolated case, and then to favor lies if the benefits seem to outweigh the costs. In so doing, they risk blinding themselves to the effects that such lying can have on their integrity and self-respect, and to the jeopardy in which they place others. (Bok 1999: xix)

There is little doubt that lying might allow an economist to achieve some immediate prosocial objective. The benefit is instant and tangible; the cost deferred, distant, and intangible. The threat to trust is all the greater when lying is tolerated and even normalized within a profession and becomes a regular rather than an exceptional feature of that profession's practice.[17] Unfortunately, the loss of trust harms not just the profession that has lost trust, but also, and perhaps even primarily, the public that may indeed need the services that the profession can uniquely provide. Moreover, a practice of deception in economics may in turn contribute to the broader loss of trust that is a central feature of the illiberal turn in contemporary politics, a central pillar of which is a deep rejection of expertise. In a world of recurrent and even escalating crises, such as the COVID-19 pandemic and climate change,

[16] One example should suffice. In response to the new empirical findings in the 1990s that suggested that raising the minimum wage sometimes increased employment, James Buchanan had this to say: "no self-respecting economist would claim that increases in the minimum wage increase employment. Such a claim, if seriously advanced, becomes equivalent to a denial that there is even minimal scientific content in economics, and that, in consequence, economists can do nothing but write as advocates for ideological interests. *Fortunately, only a handful of economists are willing to throw over the teaching of two centuries; we have not yet become a bevy of camp-following whores*" (1996; emphasis added).

[17] Those who share my worry about the damage from lying should consider Lyall's argument (Chapter 2, this volume) in favor of scholars pre-registering their ethical commitments, publicly, prior to engagement in fieldwork among vulnerable populations. Pre-commitments of this sort might go some way toward parrying the pressures that arise later on, in the research and advising stages, to engage in prosocial deception—especially in the kind of case described by Olsen (Chapter 3, this volume).

the rejection of expertise is proving to be catastrophic. Winning back trust might require a recalibration of how economists engage the world—a recalibration that enshrines honesty and respect for the autonomy of those the profession seeks to serve.

References

Ahamed, Liaquat (2009) *Lords of Finance: 1929, the Great Depression and the Bankers Who Broke the World*. New York: Penguin.

Allen, William R. (1977) "Economics, Economists, and Economic Policy: Modern American Experiences." *History of Political Economy* 9 (1): 48–88.

Bernanke, Ben S. (2008) "Risk Management in Financial Institutions. Speech given at the Federal Reserve Bank of Chicago." Annual Conference on Bank Structure and Competition, May 15, 2008.

Bernanke, Ben S. (2008) "Remarks on the Economic Outlook." International Monetary Conference, Barcelona, Spain (via satellite), June 3, 2008.

Bernanke, Ben S. (2008) "Outstanding Issues in the Analysis of Inflation." The Federal Reserve Bank of Boston's 53rd Annual Economic Conference, Chatham, MA June 9, 2008.

Bernanke, Ben S. (2008) "Regulatory Restructuring." Testimony before the Committee on Financial Services, US House of Representatives, July 10, 2008.

Bok, Sissela (1999) *Lying: Moral Choice in Public Life and Private Life*. New York: Vintage.

Buchanan, James (1996) Letter to *Wall St. Journal*, April 25, 1996.

Carrick-Hagenbarth, Jessica, and Gerald A. Epstein (2012) "Dangerous Interconnectedness: Economists' Conflicts of Interest, Ideology and Financial Crisis." *Cambridge Journal of Economics* 36(1): 43–63.

Colander, David (2019) "Are Modern Monetary Theory's Lies 'Plausible Lies'?" *Real-World Economic Review* 89: 62–71.

Colander, David, and Craig Freedman (2019) *Where Economics Went Wrong: Chicago's Abandonment of Classical Liberalism*. Princeton, NJ: Princeton University.

DeMartino, George F. (2011) *The Economist's Oath*. New York: Oxford University Press.

DeMartino, George F. (2013) "Epistemic Aspects of Economic Practice and the Need for Professional Economic Ethics." *Review of Social Economy 71*(2): 166–186.

DeMartino, George F. (2019) "Reconstructing Globalization in an Illiberal Era." *Ethics and International Affairs* 32 (3): 361. DOI: 10.1017/S0892679418000515.

DeMartino, George F. (2022) *The Tragic Science: How Economists Cause Harm (Even as They Aspire To Do Good)*. Chicago: University of Chicago Press.

Dworkin, Gerald (2005) "Paternalism." *Stanford Encyclopedia of Philosophy*, December 20, 2005. http://plato.stanford.edu/entries/paternalism/.

Flitter, Emily, Kristina Cooke, and Pedro da Costa (2010) "Special Report: For Some Professors, Disclosure Is Academic." *Reuters*. http://www.reuters.com/article/2010/12/20/us-academics-conflicts-idUSTRE6BJ3LF20101220.

Freeman, Alan (2016) "First Tell No Untruth." In *The Oxford Handbook of Professional Economic Ethics*, ed. G. F. DeMartino and D. N. McCloskey. New York: Oxford University Press, pp. 651–667.

Galtung, Johan (1969) "Violence, Peace, and Peace Research." *Journal of Peace Research* 6(3): 167–191.

Grant, Ruth W. (2012) *Strings Attached: Untangling the Ethics of Incentives*. New York City: Russell Sage Foundation.

Haldane, Andrew, Alistair Macaulay, and Michael McMahon (2020) "The 3 E's of Central Bank Communication with the Public." Staff Working Paper No. 847, Bank of England. https://www.bankofengland.co.uk/working-paper/2020/the-3-es-of-central-bank-communication-with-the-public.

Hardwig, John (1994) "Toward an Ethic of Expertise." In *Professional Ethics and Social Responsibility*, ed. D. E. Wueste. Lanham, MD: Rowman and Littlefield, pp. 83–101.

Herszenhorn, David M. and Steven R. Weisman (2008) "Republican Leader in US House Confident in Rescue Plan." *The New York Times*, July 17, 2008.

Ip, Greg and Sudeep Reddy (2008) "Bernanke Unshaken by Jobless Rise: Fed's Outlook Intact as Other Indicators Ease the Worst Fears." *Wall Street Journal*, June 10, 2008.

Kaletsky, Anatole (2014) "A Central Banker's License to Lie." *Reuters*, January 30.

Kliesen, Kevin L., Brian Levine, and Christopher J. Waller (2019) "Gauging Market Responses to Monetary Policy Communication." *Federal Reserve Bank of St. Louis Review*, Second Quarter, 101(2): 69–91. https://doi.org/10.20955/r.101.69-91.

Kool, Clemens J. M. and Daniel L. Thornton (2012) "How Effective Is Central Bank Forward Guidance?" Working Paper 2012-063A. Federal Reserve Bank of St. Louis. http://research.stlouisfed.org/wp/2012/2012-063.pdf.

Leamer, Edward E. (1983) "Let's Take the Con Out of Econometrics." *The American Economic Review* 73(1): 31–43.

Levine, Emma E., and Maurice E. Schweitzer (2014) "Are Liars Ethical? On the Tension Between Benevolence and Honesty." *Journal of Experimental Social Psychology*. http://dx.doi.org/10.1016/j.jesp.2014.03.005.

Levine, Emily E., and Maurice E. Schweitzer (2015) "Prosocial Lies: When Deception Breeds Trust." *Organizational Behavior and Human Decision Processes* 126: 88–106.

Lupoli, Matthew J., Emily E. Levine, and Adam E. Greenberg (2018) "Paternalistic Lies." *Organizational Behavior and Human Decision Processes* 146: 31–50. https://doi.org/10.1016/j.obhdp.2018.01.001.

Mayer, Thomas (2016) "Honesty and Integrity in Econometrics." In *The Oxford Handbook of Professional Economic Ethics*, ed. G.F. DeMartino and D. N. McCloskey. New York: Oxford University Press, pp. 329–351.

Mill, John Stuart (1859/2001). *On Liberty*. Kitchener, Ontario: Batoche Books.

Robinson, Joan (1978) "Marx, Marshall, and Keynes." *Occasional Papers No. 9*, Delhi School of Economics.

Rodrik, Dani (2018) *Straight Talk on Trade*. Princeton, NJ: Princeton University Press.

Roth, Alvin E. (1994) "Let's Keep the Con out of Experimental Econ: A Methodological Note." *Empirical Economics* 19(2): 279–289.

Sartre, Jean-Paul (1960) "Dirty Hands." In *No Exit and Three Other Plays*, trans. Lionel Abel. New York: Vintage, pp. 125–241.

Sharpe, Virginia A. and Alan I. Faden (1998) *Medical Harm: Historical, Conceptual, and Ethical Dimensions of Iatrogenic Illness*. Cambridge: Cambridge University Press.

Shuster, Evelyne (1998) "The Nuremberg Code: Hippocratic Ethics and Human Right." *Lancet* 351(9107): 974–977.

Stiglitz, Joseph (2016) "Ethics, Economic Advice, and Economic Policy." In *The Oxford Handbook of Professional Economic Ethics*, ed. G. F. DeMartino and D. N. McCloskey. New York: Oxford University Press, pp. 495–519.

Stiglitz, Joseph (2018) *Globalization and its Discontents Revisited*. New York: W. W. Norton.

Thompson, Dennis F. (1987) *Political Ethics and Public Policy*. Cambridge, MA: Harvard University Press.

Veatch, Robert M. 1984. "Autonomy's Temporary Triumph." Hastings Center Report. (October): 38–42.

Walzer, Michael (1973) "Political Action: The Problem of Dirty Hands." *Philosophy & Public Affairs* 2(2; Winter): 160–180.

Wilson, Bart (2016) "The Meaning of 'Deceive' in Experimental Economic Science." In *The Oxford Handbook of Professional Economic Ethics*, ed. G. F. DeMartino and D. N. McCloskey. New York: Oxford University Press, pp. 317–326.

Winston, Kenneth I. (1994) "Necessity and Choice in Political Ethics: Varieties of Dirty Hands." In *Professional Ethics and Social Responsibility*, ed. Daniel E. Wueste. Lanham, MD: Rowman and Littlefiled, pp. 37–66.

Wolpe, Paul R. (1998) "The Triumph of Autonomy in American Bioethics: A Sociological Approach." In *Bioethics and Society*, ed. R. DeVries and J. Subedi. Upper Saddle River, NJ: Prentice Hall, pp. 38–59.

Wueste, Daniel E., ed. (1994) *Professional Ethics and Social Responsibility*, Lanham, MD: Rowman and Littlefield.

7

The Problem of Overgeneralization

The Effects of NATO Enlargement

Rachel A. Epstein

This chapter examines the extent to which researchers writing about the North Atlantic Treaty Organization's (NATO) enlargement to postcommunist countries from 1999 onward "overgeneralized" their findings.[1] That is to say, some researchers misapplied earlier findings prospectively or made causal claims that were unsupported by evidence. I also consider the related problem of policymakers selectively seizing on only some aspects of scholarly results, leaving aside academic circumspection in favor of conclusions they preferred. This review of the NATO enlargement debate and the academic research that informed it is intended to help scholars—no matter their field—be aware of overgeneralization and avoid it. Scholars should further be cognizant of policymakers' possible mischaracterization of research findings in their efforts to advance a policy position. In addition, this chapter pinpoints the ways in which unacknowledged values shaped the NATO enlargement debate. Although researchers and policymakers alike were putatively arguing over evidence, in fact underlying but unspoken political commitments were at stake as well.

NATO enlargement has had far-reaching effects, though more than thirty years following the Cold War's end there is little agreement over what those effects have been. In the intervening decades, Russia has become increasingly authoritarian, repressive, and violent. It has incited and supported separatist movements in Georgia and Ukraine, eradicated separatist movements in Chechnya, provided extensive military assistance to Bashar al-Assad in Syria, annexed Crimea, and launched a full-scale invasion of Ukraine in February 2022, using Belarus as a major staging ground, therewith

[1] I would like to thank Anna Kulikova, Chuck Aoki, Anushka Bose and Louis Pascarella for their research assistance on this project. I would also like to thank Alex Downes, Cullen Hendrix, Oliver Kaplan, Martin Rhodes, the Robert Schuman Centre seminar series at the European University Institute, and the "Contesting the Liberal Script" seminar series at Freie Universität in Berlin (and especially Tanja Börzel and Thomas Risse) for their comments on earlier drafts. All errors are my own.

Rachel A. Epstein, *The Problem of Overgeneralization*. In: *Speaking Science to Power*. Edited by: Rachel A. Epstein and Oliver Kaplan, Oxford University Press. © Rachel A. Epstein (2024). DOI: 10.1093/oso/9780198875185.003.0007

cementing that country's client status in the process. Meanwhile, by the time of writing, NATO had enlarged from sixteen to thirty-two members and had engaged in out-of-area military operations in Europe, the Middle East, and Central Asia. It also coordinated a unified Western response to Russia's 2022 invasion of Ukraine that included invitations for Sweden and Finland to join the alliance. Democratic consolidation across NATO's postcommunist members (not to mention Turkey) was uneven, while regional rivalries among states within the alliance had been contained. These developments were not in dispute, though NATO's role in causing them was.

Establishing causality in social science is challenging. This creates a dilemma for researchers seeking to engage in policy debates. Many observers believe policymaking could be improved if it were predicated on sound findings. Yet findings are never so sound that there are no risks in implementation. This is not only because research has its limits, but also because one cannot anticipate upending events and hidden variables that make one setting unexpectedly different from another. Reviewing the NATO debate, I find that scholars were more likely to err on the side of *overstatement* rather than caution, in some cases predicting Russia's radicalization in response to NATO enlargement when any number of factors could lead to that outcome. While this may be understandable from the perspective of being heard and having influence, researchers were not always aware of (or at least did not state) the extent to which their claims and related policy recommendations exceeded their evidentiary foundations, relied on partial or inapt historical analogies, or stemmed from theories that embed biases (see also DeMartino, Chapter 6, this volume). I argue that researchers can take additional steps to improve the research–policymaking interface by sharpening the critical lens through which they present their findings.

In response to the problem of overgeneralization, I suggest seven strategies for mitigating the unintended consequences of overgeneralization. The first three are deeply embedded in most PhD training and are simply a repetition of best practice. These include, first, thoroughly investigating alternative explanations; second, articulating the limits of any given study; and third, outlining how unpredictable future events may undermine the salience of a study's causal or constitutive variables. Research methodology, while not the focus of this chapter, is clearly critical to the robustness of any claim (see also Lyall, Chapter 2, this volume).

Additional suggestions for avoiding overgeneralization by scholars come into conflict with some advanced training. For example, the fourth strategy is for researchers to be explicit about their values and how those affect their conclusions rather than claim that particular methodological and theoretical choices are unreflective of prior political commitments (Zvobgo and Loken 2020; Gani and Marshall 2022). Though scholars might believe their research is objective, the case of NATO shows how particular unacknowledged values seeped into the analysis and conclusions. Fifth, despite the drive for parsimony, clarity, and policy influence that one is more likely to achieve with stridency than with caution (see also Barma and Goldgeier 2022), academics should refrain from overstating their findings by being explicit about the evidentiary basis of their claims and the degrees of uncertainty, and by critically examining the historical analogies they deploy. Even in op-ed format, alerting readers to limits and counterarguments is essential. Sixth, and on the other side of that equation, policymakers would ideally embrace the complexity, contingency, and uncertainty that are intrinsic to any research—and explain it to their audiences. Seventh, and finally, analysts could periodically review their past work, critically evaluating the quality of their research and findings in retrospect, their tendencies toward overgeneralization, and the robustness of their policy recommendations or actions. Continual reflection and forthrightness about past errors are crucial to constructive policy contributions.

How NATO Enlargement Illustrates the Problem of Overgeneralization

NATO enlargement and its effects on security, conflict, and democratic development are useful for exploring the problem of overgeneralization because, from the 1990s onward, scholars and policymakers have examined the same evidence and come to radically different conclusions about the merits of the policy (see also Sisk, Chapter 4, this volume, on competing claims). Russia's seizure of Crimea, its fomenting of conflict in Eastern Ukraine, and its full-scale invasion of the country in February 2022—alongside NATO's continuing enlargement—have reinvigorated the debate over enlargement.[2] Starting in the 1990s, both researchers and policymakers made a series of strong claims on behalf of one side or the other, pointing variously to the risk of Russian revanchism, the need to mitigate regional rivalries, and the opportunity to consolidate democracy. With the benefit of more than two decades of evidence, we can adjudicate among scholarly findings, pinpointing those to which scholars and policymakers attributed too much or too little

[2] See *Foreign Affairs*, "Ask the Experts," April 19, 2022: https://www.foreignaffairs.com/ask-the-experts/2022-04-19/was-nato-enlargement-mistake.

weight. The literature also shows the underlying values of those carrying out research and policy, which not only affected how findings were framed, but also led to ongoing disagreement about what the evidence says (Goldgeier and Shifrinson 2020).[3]

The debate over NATO enlargement is also useful for exploring the problem of overgeneralization because it is a clear case of policy importance in which the harmful consequences of getting the policy wrong are significant. Thus, specifying arguments carefully is critical. Since the end of the Cold War, when NATO might have folded and alternative security arrangements were potentially on offer, the alliance has admitted no fewer than *sixteen* new countries, doubling its membership since 1999. If this arrangement, as critics argued, fomented nationalism in Russia, provided adequate justification for Russia to invade its neighbors, prevented consensus within NATO by virtue of too many conflicting political ideologies and interests, or compromised military readiness because NATO's Article V security guarantee encouraged countries to "cheap ride" on US military investments (Goldgeier and Shifrinson 2020), policymakers need to know and change course. This call for assessment is equally urgent if, conversely, enlargement advanced democratic governance, contained authoritarian powers, mitigated regional rivalries, and increased Western military and political capacity. Through identifying and then learning to curb overgeneralizations, researchers might limit their enabling of misguided policy. While this chapter is focused on the NATO enlargement debate, it invites scholars analyzing other issues in international affairs to consider the extent to which overgeneralization infuses policy dilemmas more broadly.

The Early Debate over NATO Enlargement

As a party to the debate over NATO enlargement, I am not a neutral observer. I argued that although NATO's effects were uneven, on balance its enlargement had been democratizing and denationalizing for postcommunist countries (Epstein 2005, 2006, 2008).[4] Having started my research in the mid-1990s as an enlargement skeptic, fieldwork in East Central Europe (ECE) for

[3] Their special issue of *International Politics* encapsulates ongoing debates, including whether NATO has supported democratic development, whether adding new members helps or hinders the alliance's functioning, and whether NATO enlargement has contributed to or detracted from relative peace and stability in the post–Cold War period, particularly on the Eastern side of the European continent. Goldgeier and Shifrinson also point to the fact that values, not just evidence, inform their contributors' conclusions, and those of others in the debate (pp. 313–314).

[4] See also my blog post for *Political Violence @ A Glance* on the subject from 2016: https://politicalviolenceataglance.org/2016/09/13/why-nato-enlargement-was-a-good-idea/.

much of 1999 changed my mind. In addition to finding democratizing and denationalizing effects among the newest members, my political convictions increasingly turned toward supporting democratic sovereignty—meaning that countries long-subjugated to large powers' priorities and negotiations should be able to decide for themselves with whom they wanted to ally. From this political perspective, both NATO and EU enlargement offered Poland, Hungary, the Czech Republic and a host of other newly independent countries membership, voice, and voting rights within two of the world's most powerful multilateral organizations (also see Vachudova 2005).

In my view, the open-door policies of the EU and NATO represented a partial upending of the normally strictly hierarchical conduct of international politics. The initial requests of East Europeans to join, the readiness of the EU and NATO to prioritize these newly independent but small countries' interests, and the ultimate decision to invite them in appeared to acknowledge historical prejudicial injustices. Those injustices, perpetrated by so-called "great powers," had earlier included Munich, the Katyn massacre, the "percentages agreement," the abrogation of Yalta, and the Soviet takeover of Eastern Europe following World War II.[5] By the time I had completed my field research, I concluded that while NATO enlargement could never undo the damage wrought by large powers using smaller ones as pawns, at least East European populations were entering an era in which they would have more discretion over their own affairs. Though I held these views, I never stated them explicitly in my academic work. I was trained to evaluate evidence, not values.

In reviewing the original literature from the early days of the enlargement debate, I find statements of possible overgeneralization and consider the evidentiary foundations on which I and other scholars made those claims. Although I do assess those early assertions according to subsequent empirical developments, my primary aim in this chapter is *not* to adjudicate the debate about NATO enlargement. Rather, the goal is to evaluate whether scholars engaged responsibly in the policy debate or whether they overreached, failed to explore the full range of alternative explanations, used inapt historical analogies to persuade, were not explicit about their normative commitments when they had them, or could have done more to specify the conditions under

[5] "Munich" is shorthand here, not primarily for appeasement, but for the negotiation away of Czechoslovak prerogatives by more powerful states at the time (Albright 2012). The Katyn massacre refers to the Soviet murder of more than 20,000 Polish military officers during World War II, for which it was never held legally accountable. The "percentages agreement" was an informal understanding between Churchill and Stalin about spheres of influence in Europe, to the ultimate detriment of Eastern Europe. The abrogation of Yalta refers to the failure to defend democratic elections for the peoples of Europe (Rothschild and Wingfield 2020).

which alternative scenarios might have materialized. To foreshadow, from both academic and policy perspectives, we could have done better.

Scholars who argued that NATO had democratizing, denationalizing, and stabilizing effects through enlargement drew on a range of indicators. They found that, vis-à-vis ECE, NATO mitigated regional rivalries and mistrust, socialized actors into embracing the desirability of democratic civilian oversight, and discouraged relative defense self-sufficiency in favor of professionalized and expeditionary forces (Wallander 2000; Moore 2002; Flockhart 2004; Jacoby 2004; Epstein 2005, 2008; Gheciu 2005a, 2005b; Lucarelli 2005). Evidence for these assertions was observed in the comparison of actors' beliefs before and after contact with NATO, the translation of those beliefs into legislation and defense reform consistent with NATO prescriptions, the discontinuation of direct military interference in politics, the civilianization of ministries of defense, the creation of parliamentary committees charged with military oversight, the increasing transparency of defense budgets, the signing of peace treaties with neighboring countries, the affirmation of borders and national belonging (including of diaspora populations), the curtailment of territorial or "total" defense, the downsizing of militaries everywhere except the Baltics (where national forces had to be constructed anew), and the creation of expeditionary forces for out-of-area missions—ultimately bound for Afghanistan and Iraq. Overall, this literature was careful to specify NATO's limits, as well as the reasons for variation in the alliance's reform effects.

The extensive list of reforms outlined above led me in 2005 to conclude that NATO's policies had begun to "internationalize the orientation of states and societies" in ECE (Epstein 2005: 65–66). I also argued that NATO had "precluded the rise of destructive military cultures by insisting on democratic standards" in the region and that "the pessimistic predictions surrounding NATO enlargement never materialized" (65), implying, arguably prematurely, that they never would. In direct repudiation of skeptics, I maintained that NATO had "contributed to democratization, as well as to other positive trends among its member states," including having "significantly denationalized defense strategies . . . thereby stabiliz[ing] relations among states" (63–64). The alternative explanations I explored were less about the empirical results or possible future trajectories than about *how* NATO and its domestic coalition partners achieved reform. Whereas some scholars emphasized the role of material incentives and conditionality, others (including myself) argued that socialization, social learning, in-group dynamics centered on perceived civilizational standards and shaming were more important.

On the other side of the debate, critics of NATO enlargement argued that the alliance did not have a history of aiding democratic consolidation and was not equipped to do so in the post–Cold War period. Democratization, they concluded, was either a matter of domestic politics alone or would be better served by EU conditionality and membership predicated on reform. They also warned that NATO enlargement exploited Russian weakness in ways that were not only unfair, but that also risked inciting Russian nationalism to the detriment of Russian–Western relations, possibly creating incentives for a Chinese–Russian alliance. Moreover, too wide an array of states in NATO threatened alliance cohesion (particularly as decision-making was consensus-based) and undermined the credibility of NATO's Article V security guarantee by incorporating geographically exposed countries near Russia, even as NATO's own logistics were wanting (Mandelbaum 1995; Kennan 1997; Eisenhower 1998; Gaddis 1998; Russett and Stam 1998; Reiter 2001).[6]

Revisiting some specific statements prior to the first postcommunist enlargement, Mandelbaum (1995: 9) argued that "NATO is not an effective instrument for promoting either free markets or democracy" and that, with regard to East Central European countries' internal reforms or security challenges, the policy of enlargement was "largely irrelevant to the problems confronting countries situated between Germany and Russia." Further, on the geostrategic consequences of enlarging NATO (or not), Mandelbaum claimed that "an independent Ukraine is the best guarantee that Russia will remain a peaceful nation-state" but that an enlargement policy that did not include Ukraine (however unwise enlargement would have been in all other respects) "would in all likelihood weaken its [Ukraine's] commitment to relinquishing to Russia all the nuclear weapons it inherited from the Soviet Union" (1995: 10).

Reiter, another critic of NATO enlargement, agreed that NATO had no democratizing influence and that "the West should rely on the EU to spread democracy" (2001: 42). Somewhat different from Mandelbaum and at odds with George Kennan, who predicted that enlargement would "inflame the nationalistic, anti-Western and militaristic tendencies in Russian opinion" (Kennan 1997: fn 6), Reiter made a different argument against enlargement. Reiter's argument hinged not on the fear of inflaming Russian nationalism, but rather on the idea that enlargement was not necessary because the "prospect of renewed Russian aggression against any of the current NATO

[6] George Kennan's opinion piece on NATO enlargement argues that the policy was a "fateful error": https://www.nytimes.com/1997/02/05/opinion/a-fateful-error.html. See also Thomas L. Friedman's piece from the same period: https://www.nytimes.com/1997/01/22/opinion/nato-or-tomato.html.

applicants is extremely remote" (46). Reiter cited the "sheer improbability" of the revival of Russian nationalism or military capability worthy of concern (47). Perhaps also prematurely, Reiter observed in 2001 that "Russia has exhibited no inclination to rebuild the Soviet empire through threats of force" (46), without acknowledging this could change in the long or even medium term.

Some clear and competing claims emerge from this summary. While many of the findings were valid from the point in time in which they were made, in some cases scholars overgeneralized from those points in time—assuming they would apply in the future as well. Policymakers likewise overstated their cases, often before systematic research was available. Moreover, a wider consideration of alternative explanations, limitations of the research, and possible future scenarios was in order. Normative commitments made their way into the literature unacknowledged. In what follows, I highlight claims that could not be supported by evidence, that were made despite conflicting evidence, or that were predicated on undisclosed predispositions. To identify these weaknesses, I draw on more recent findings and unexpected outcomes.

The Effects of NATO Enlargement: Evaluating Claims

I group the claims that structured the debate about NATO enlargement and its effects into three areas. The first is that NATO enlargement would antagonize Russia, justify that country's nationalist proclivities or outward aggression, and lead to a downward spiral in Russian–Western relations. The second is that NATO denationalized defense and foreign policy by enforcing peaceful conduct among members and neighbors, mitigating regional rivalries, and reducing the potential for conflict within the transatlantic community. The third is that NATO enlargement would advance and support democratization by introducing a new model of democratic civil–military relations to erstwhile communist countries or to newly independent countries that had formerly been part of the Soviet Union or Yugoslavia. In some cases, I find overgeneralization. In others, there are unacknowledged political commitments or underexplored alternative explanations.

NATO Enlargement and Russian Revanchism: Overgeneralization and Unacknowledged Values

Early arguments against NATO enlargement to ECE included that it would incite Russian nationalism, elevate that country's sense of humiliation, and lead to the dramatic worsening of Russia–Western relations (Mandelbaum 1995; Kennan 1997; Eisenhower 1998). Relatedly, some scholars argued that NATO enlargement would not only incite Russian nationalism, but would also push Russia into an alliance with China. For some, therefore, the only way to proceed with enlargement was to include Russia (Gaddis 1998; Russett and Stam 1998). On the other side of the revanchism debate (but still critical of the enlargement policy), scholars downplayed Russia's future capacities or potential for imperial ambitions (Reiter 2001).

In one way, Kennan's very short essay calling the prospect of enlargement a "fateful error" appears prescient, drawing attention as he did to the likelihood of democratic failure in Russia and Russian foreign policies at odds with Western interests (Kennan 1997: fn 6). But even with the benefit of hindsight, we still cannot know whether NATO enlargement *caused* Russian aggression, revanchism, and heightened hostility with the West, though recent insider accounts cast doubt on the assertion (Bondarev 2022). But the unknowable nature of the claim did not stop other observers from expressing strong opinions on the subject. According to Russett and Stam in 1998, Russia's perception of the NATO threat risked "reviving old Russian fears of the West, strengthening Russian militarists and nationalists, and inducing greater instability in Russian domestic politics and foreign policy" (1998: 362). In fact, this has been Russia's trajectory. But the statement and broader article did not consider the counterfactual possibility that Russian politics might have evolved similarly even *without* NATO enlargement.[7]

With access to additional documents and attention to the larger political context, Sarotte has more recently concluded that the policy of enlargement "did not by itself spell the demise of US–Russian post-Cold War cooperation" (2019: 10). She points to grave errors on both sides of the Russian–Western relationship—including the Russian crackdown in Chechnya but also the self-serving economic policy advising by Westerners. In doing so, she highlights that while NATO enlargement might have been structured more cautiously, additional variables bore even more forcefully on the relationship. Some early (and even some later) observers ignored this fact (see especially Mearsheimer 2014, 2022).

When it comes to whether NATO enlargement fomented Russian nationalism and aggression, scholars on both sides of the debate can marshal confirmatory evidence for their positions (Figure 7.1). More important, however, is how this debate reveals the subtle ways in which normative

[7] See the *New Yorker* interview with Stephen Kotkin, a historian of Russian politics, who dismisses the idea that Russia's invasion of Ukraine was either historically inconsistent with past Russian conduct or primarily a response to Western policy. "The Weakness of the Despot," March 11, 2022.

Figure 7.1 NATO Secretary General, Jaap de Hoop Scheffer, and the President of Romania, Traian Basescu, unveil the official logo for the 2008 NATO summit in Bucharest. NATO would decide on an open-door policy, inviting Albania and Croatia to accession talks, welcoming Georgia and Ukraine's "aspirations for membership," and inviting Bosnia and Herzegovina and Montenegro to "Intensified Dialogue."
© NATO

commitments made their way into research. While such commitments were never clearly articulated or even acknowledged in the NATO debate, they influenced findings (Goldgeier and Shifrinson 2020).[8]

In response to the above arguments about NATO inciting Russian nationalism, supporters of NATO enlargement offered the following observations: That successive Russian regimes were riddled with corruption and intent on covering it up, including by violent means. From this perspective, Russian leaders would periodically use NATO enlargement as a distraction from domestic policy failures and as justification for helping to quash or directly repress democratic protest movements in neighboring countries. Putin's misrepresentations of early Western–Russian agreements that he falsely claimed included commitments prohibiting NATO enlargement provide evidence

[8] This point was also made by Daniel Nexon in the *Foreign Affairs* survey "Ask the Experts" from 2022, fn 2. Though he says the debate is a proxy for other disputes, in his very short explanation he does not say what those disputes are.

of Putin's disingenuous deployment of a NATO threat (see Sarotte 2019, 2021 on the absence of any such commitment). Moreover, advocates of NATO enlargement contended that governments from Yeltsin to Putin very likely massacred their own citizens through a series of domestic apartment-building bombings (blamed on Chechnyan separatists) to build support for authoritarianism (Satter 2016). Russia also consistently pursued imperialist ambitions, regardless of the will or sovereignty of neighboring states (Grigas 2016). Whether NATO enlarged or not, this argument went, democratic endurance in Russia was never likely. And, given Russian and Central-East European history, it was better to secure the small, vulnerable states between Russia and Germany than to risk further curbs on their autonomy—or worse—by Russia (Lanoszka 2020).

But beneath the surface, another debate was playing out. Some scholars prioritized relations between major powers (sometimes referred to as "great powers"), particularly Russia and the US in this case. Others, however, were more concerned about democratic sovereignty and advancing the rights, prosperity, and security of the countries in "Zwischeneuropa"—those states sandwiched between Russia and Germany that had long been beholden to large power rivalry. Structural realists and "great power" historians implicitly and sometimes explicitly argue that large powers take priority. "Great power" prioritization assumes two forms. The first is that these scholars believe big countries should command more of our attention in terms of study. But, second, they argue that powerful countries are justified in taking more power when they can get away with it, even if it comes at the expense of others (Mearsheimer 2018).

There are two values debates at stake. The first is over democratic sovereignty versus the claim that "great powers" are justified in using force to maximize their power. The second is over who has voice. By focusing on large powers disproportionately, the perspectives of those countries are overrepresented relative to other actors. Moreover, while realists will claim that their theory is value neutral (in contrast to, for example, liberal idealists or democratic sovereigntists), the pretense to scientific credibility is dubious. There is nothing value neutral in arguing that large powers are justified in power maximization, including by violent means.

On the side of privileging "great powers"—a term I do not use because it connotes that some countries are not only larger than others but also superior in some sense—Stephen Cohen, for example, argued that after the break-up of the Soviet Union in 1991, "[T]he foremost goal of US policymaking should have been a Russia ... that was prospering, politically stable, at peace and fully cooperating with the US on the most threatening international problems"

(2001, p. 9). John Lewis Gaddis agreed, stating that among the most important rules of foreign strategy was to "treat former enemies magnanimously" and that NATO enlargement violated not only this basic diplomatic principle, but every other as well (quoted in Reiter 2001: 48). Mandelbaum's assertion that the policy of enlargement was "largely irrelevant to the problems confronting countries situated between Germany and Russia" (1995: 9) mirrored another sentiment in the literature: that East Central Europe was not in the West's strategic interest to protect. As Reiter matter-of-factly pointed out as justification for why NATO should not enlarge to ECE, the "West did after all accept Soviet annexation or domination of all of these states during the Cold War without taking military action" (2001: 48).

For specialists working primarily on ECE, however, the statements above are ethically problematic, even if there are few empirical disagreements. For example, many scholars of ECE would agree on the goal of constructive Western relations with Russia but would not agree that such relations should come at the expense of the East's ongoing exposure to possible Russian intervention, violence, and revisionism. Second, there is no empirical disagreement over Western and US abdication of Eastern Europe following World War II—though many analysts of the region do contend that it was a catastrophic mistake, not to be repeated (Rothschild and Wingfield 2000). Third, and finally, proponents of NATO enlargement are continually asking their critics whether the sentiments of those "situated between Germany and Russia" should matter in this controversy. Defenders of the enlargement policy argue that from a democratic sovereignty perspective, the world should listen to those populations and leaders who had not willingly succumbed to Soviet domination. Many would argue further that this sentiment should not be confined to Europe but is a universal value that should apply to formerly and currently subjugated peoples everywhere. And here, there is an empirical correction—research shows that the drive for NATO enlargement originated within postcommunist countries—so, by the East Europeans' own assessment and against Mandelbaum, the policy of enlargement was certainly *not* "irrelevant" to the problems they confronted (Goldgeier 1999; Epstein 2008; Sarotte 2019).[9]

The values debate went even deeper, however, with at least one critic of NATO enlargement accusing supporters of the policy, including those within the Clinton administration, of "selective sentimentalism" (Gaddis 1998: 149).

[9] See also the Deutsche Welle interview with Horst Teltschik from 10 February 2022, long-time German foreign policy advisor, including to Chancellor Helmut Kohl, "Kohl-Berater Teltschik: Es hat von Deutschland nie ein Versprechen über die Nato-Erweiterung gegeben": https://www.youtube.com/watch?v=NnVVpsCH_AE.

Reminiscent of the Realist E. H. Carr, who famously argued that the world for which liberal idealists hoped departed dramatically from the world as it actually is (Carr 1945), Gaddis argued that East Central European suffering in the twentieth century had been no more severe than for the Russians themselves. They were all, in Gaddis's words, "in one way or another, victims of German and/or Soviet oppression" (Gaddis 1998: 149). The fundamental mistake behind the NATO enlargement policy, according to Gaddis, was "emotionalism" on the part of three key US officials toward East Central European "suffering," acted upon without adequate debate or input from the relevant arms of government, including the US Department of Defense (Gaddis 1998: 149, 145).

While many experts of the region would point out that equating all victimhood is itself an overgeneralization that scholars could debate at length, there is a more fundamental disagreement about values in Gaddis's writing. Because realists in International Relations have long argued that states' survival, power positions, and competitiveness dictate conduct in international affairs, they believe they can draw a clear distinction between what is and what should be. This is a common expression of research objectivity at the advanced level of social science training, which some social scientists further claim exempts them from "emotionalism" and any other form of political bias. And yet the theorized transcendence of self-regarding state conduct in International Relations is also an overgeneralization that ignores the extent to which our participation in the world reinforces a status quo or transforms it (Zvobgo and Loken 2020). Insisting on the primacy of large powers' interests is not politically neutral, just as the enfranchisement of smaller powers' interests is not either. In this instance, Gaddis's drive for objectivity and rejection of "emotionalism" also led him to draw what were arguably inapt historical comparisons: between Germany at the end of World War II and Russia following the Cold War. While they were both defeated powers, the conditions surrounding their reintegration into global politics were entirely different, calling into question his argument that as few as six basic principles always apply (Gaddis 1998: 145–146).

Further evidence of theoretically embedded bias leading to overgeneralization comes again from John Mearsheimer, who also blamed elite idealists in Washington, DC for Russia's invasion of Ukraine. Writing about Russia's annexation of Crimea in 2014, Mearsheimer argued that "Putin's actions should be easy to comprehend" since Ukraine "serves as a buffer state of enormous strategic importance to Russia" (2014: 82). He advances the idea of a clear hierarchy of states in which some have more rights than others, a value that he does not acknowledge. He also claimed that Putin is "a first class strategist who should be feared and respected" and who "surely understands that trying to subdue Ukraine would be like swallowing a porcupine" (85), meaning that Mearsheimer believed, even after the seizure of Crimea, that there was no way Putin would lead a full-scale invasion of Ukraine. Not only could Mearsheimer not have known that at the time of writing, but of course he turned out to be entirely wrong about Putin's strategic acumen and the scale of the invasion. In addition, he provided no direct evidence from within the Russian policymaking apparatus about the underlying motivations for the invasion, arguing instead that we should take Putin at his word that Western aggression was the provocation.

Unacknowledged bias and the temptation to overgeneralize may stem from researchers internalizing the values and interests of the countries and populations they study, depending on the empirics they absorb, the history they read, and the field research they do. Being aware of these influences, as well as one's own power (or lack thereof) vis-à-vis research subjects is crucial to understanding not just the evidentiary but also the social sources of argumentation (see also Kaplan, Chapter 5, this volume). This kind of socialization should not necessarily be discouraged since it can lead to sharp insight. For instance, Valerie Bunce, long more immersed in the politics, economics, and languages of ECE than her Sovietologist counterparts, identified fissures in the Soviet empire that presaged its collapse, even before Gorbachev came to power (Bunce 1985). Part of what distinguished Bunce's work in the 1980s was her understanding and articulation of the Soviet Union's ideological failure and its perceived illegitimacy and hypocrisy in the broader region—in other words, an understanding of the emotional context informed by disgust (see also the Introduction to this volume). This was in addition to the exorbitant costs the Soviet core was paying to sustain the empire, through energy subsidies, debt liabilities, and the like.

Scholars could acknowledge their political positions outright and clarify for policymakers where empirical differences end and value judgments begin to avoid the problem of "stealth issue advocacy," whereby researchers hide their values behind science (Pielke 2007). They could also examine more critically the political biases built into their theories and starting assumptions. This includes acknowledging that the empirics we gather and analyze are never really separate from our political priors (Stone 2018), a point that complicates much social science training and the striving for objectivity (Friedman 1953). Researchers can and hopefully always will be open to evidence changing their views. But understanding the relationship between one's interpretation of evidence and values could bring greater clarity to policy debates.

NATO's Denationalizing Effects: Overgeneralization by Policymakers

Early statements from American policymakers and research by scholars argued that NATO had "denationalizing" effects on its members. Membership encouraged cooperative security, collective defense, and/or the relocation of foreign policy and national security concerns away from the national to the supranational level (Schimmelfennig 2003; Adler 2008; Epstein 2005, 2008; Gheciu 2005a). These findings were in keeping with the first Secretary General Lord Ismay's assertion that the purpose of NATO was not only to keep the Russians out and the Americans in, but also to keep the "Germans down"—that is, to stop members from acting aggressively on their own, particularly vis-à-vis former rivals, including those in the alliance. In 1997, then-Secretary of State Madeleine Albright similarly observed that NATO had "helped former adversaries reconcile, making European unity possible. It brought the former fascist nations, first Italy, then Germany, then Spain, back into the family of European democracies. It denationalized European defense. It stabilized relations between Greece and Turkey. All without firing a shot."[10] Strobe Talbott also concluded that NATO "can foster integration and cooperation among the Central Europeans themselves," adding that "in pursuit of their goal to join NATO, a number of Central European states have accelerated their internal reforms and improved relations with each other."[11] The US Ambassador to Poland, Daniel Fried, even went so far as to say in 1999 that "NATO had solved Russia's Germany problem," meaning that by rechanneling German military power through a multinational alliance as a democratic country, NATO had also brought Russia a measure of stability and security it had not enjoyed before. Fried was arguing that the implication of NATO enlargement in 1999 was that further stability in ECE through the same means would also benefit Russia.[12]

Later research that examined NATO's effects on ECE substantiated the claims of these policymakers, except for Ambassador Fried's, whose statement of course erroneously assumed that Russia would not equate "democratic stability" on or near its borders with a threat. But those statements were made before research about the extent to which earlier cases were

[10] See the Secretary's prepared statement for the US Senate Armed Service Committee: https://1997-2001.state.gov/statements/970423.html. The fuller statement from Lord Ismay was that NATO was formed to keep the Russians out, the Americans in, and the Germans down.

[11] See coverage of the Deputy Secretary of State's remarks to the Harvard University Project on Ukrainian Security: https://1997-2001.state.gov/policy_remarks/1998/980408_talbott_ukr_nato.html.

[12] Remarks by the US Ambassador at a US–Polish Chamber of Commerce meeting in Warsaw in March 1999, attended by the author.

representative of prospective cases was available and without acknowledging possible risks. This kind of overgeneralization and the lack of thorough-going debate about NATO enlargement among policymakers and with the public was another point of criticism among skeptics (Kennan 1997: fn 6; Gaddis 1998).

What research ultimately found was that NATO insisted on the formalization of border recognition between aspiring member states and their neighbors. This resulted in a series of treaties and agreements between countries that had been former belligerents, subjects, or otherwise engaged in violent or asymmetrical relationships, such as Poland and Germany, Poland and Lithuania, Poland and Ukraine, the Czech Republic and Germany, Romania and Hungary, Romania and Ukraine, and so on. NATO also helped stabilize the status of diaspora populations and their linguistic and cultural rights, easing tensions within the region (Williams 2002). Furthermore, NATO encouraged the end of conscription, the professionalization of the armed forces, and the curtailment of territorial or "total" defense. Finally, in the downsizing of militaries, the alliance called on new members to focus on building small and specialized expeditionary forces—some of which found their way to the former Yugoslavia and ultimately to Afghanistan and Iraq in multilateral missions. In fact, Eastern aspirants to NATO were strongly discouraged from justifying their drive to join the alliance with arguments about past or potential Russian aggression (Epstein 2008). NATO and US officials were, in the 1990s and early 2000s, still eager to build a more constructive relationship with Russia, even ruling out that that country would necessarily be excluded from NATO indefinitely (Sarotte 2019).[13]

In retrospect, research findings about NATO's denationalizing effects do not appear to have been overstated. However, the empirical grounds on which policymakers' claims were made were rooted in the past, without consideration of what might change in the postcommunist context. Indeed, intervening conditions, including democratic erosion and out-of-area operations, have reversed some earlier trends, which researchers in the late 1990s and early 2000s might have explored in more depth. At the same time, the border arrangements agreed to in the run-up to NATO enlargements remain intact. None of the regional rivalries among NATO members has escalated. Countries did shift resources away from total defense to expeditionary forces, and new member states sent troops out-of-area and supported NATO's multilateral missions in other ways. Conscription was essentially wiped out across

[13] Strobe Talbott's lecture at Stanford University, 19 September 1997, available at: https://1997-2001.state.gov/regions/nis/970919talbott.html. See also the Deutsche Welle interview with Horst Teltschik, fn. 9.

the region, while defense spending hovered at or well below 2 percent of GDP in the run-up to Russia's invasion of Ukraine. Low levels of defense spending have been a source of frustration for successive US administrations, but were also a sign that NATO's European members, including in the East, did not suffer a sense of existential insecurity in a part of the world that has traditionally been extremely dangerous.

However, more critical vis-à-vis the denationalizing literature, NATO may have fallen short of "socializing states into conceptualizing their defense at the supranational level" (Epstein 2005: 93), at least in the longer term. While regional rivalries within NATO had mostly been kept at bay, territorial or total defense made a comeback, new member states questioned the value of ever having participated in multilateral missions such as Afghanistan and Iraq, and Russian aggression in Georgia and Ukraine renewed attention to collective defense within the region (Reeves 2019; von Hlatky and Fortmann 2020). To the extent that Eastern members had been willing to publicly justify NATO membership in the 1990s and early 2000s for reasons other than Russian aggression, that ended with Russia's increasing military interventions in its near abroad.

At the same time, the early literature on denationalization, socialization, cooperative security, and changing modes of defining and achieving security did not claim that the strategic context should have no effect on states' behavior. One critical observation targeting that earlier research states that the "constructivist argument would lead one to expect that the Baltic countries and Poland would need less and less reassurance over time" (von Hlatky and Fortmann 2020: 560). But this critique was divorced from the fact that constructivists also understand the reconstruing of security when an existential threat materializes, and Russia since its 2008 invasion of Georgia (if not before) was certainly that. No analyst of cooperative security ever argued that the new NATO members' "transformation should include leaving the past behind, fully embracing their new security identity as NATO member states rather than continuing to bring up historical grievances" (von Hlatky and Fortmann 2020: 560). Rather, the constructivist literature highlighted the degree to which East European interpretations of the past (including historical grievances) competed with NATO's much more benign (in the 1990s) interpretation of Russian intentions and European security (Epstein 2006, 2008). The constructivist literature argued further that if NATO's version of events prevailed, it was highly contingent, suggesting that threat perceptions could change over time. The overgeneralization of others' findings is another reminder of best practice. Treatment of alternative explanations must be accurate and contextualized in the depiction of rival arguments.

I do concede, however, that the denationalizing literature intimated that persuasion, socialization, and social learning within NATO might be unidirectional, moving militaries, political elites, and even publics toward liberal values and not the other way. This literature was attentive to nationalist and anti-Western or anti-US sentiments among some targets of persuasion, socialization (and even shaming) and documented when attitudes and policies did not change in response to NATO actions (Gheciu 2005a; 2005b). But what scholars and policymakers generally did not consider was that the alliance itself could be an agent of illiberalism.

Douglas Porch's critique of counterinsurgency (COIN), its uses and excesses through time, as well as the manipulative mythologizing around its advantages (especially by US forces in Iraq) does not address the effects of out-of-area missions on NATO's East Central European members specifically (Porch 2013), but it does suggest some avenues of possible new research. Porch traces COIN's lineage from imperial and colonial wars and demonstrates the gap between supposedly winning hearts and minds and the reality of brute force and coercion against civilian populations. He concludes that "COIN operations force democracies to compromise the very freedoms and values that they are meant to export abroad" because efforts are made "to liberate COIN from legal restraints of due process and Western political culture" (Porch 2013: 317.) Socialization and learning within NATO under such conditions would therefore not be democratic, rule-bound, or cooperative. The expanded use of private security companies (PSCs) in connection with COIN has at times also complicated civil–military relations as they undermine the very governance structures and state institutions that COIN operatives may increasingly view as "bureaucratic and unresponsive" (Porch 2013: 332, 337; also see Avant 2005).

We do not know what effect East Central European countries' participation in multilateral "forever wars" in Iraq and Afghanistan alongside other NATO members had. Enhanced readiness, equipment upgrades, professionalization, and even greater exposure to democratic norms *might* have been the consequence. But the security community and denationalization literature could have considered the alternative possibilities, including racist radicalization, given that NATO greatly expanded its out-of-area missions immediately after the 1999 enlargement, including in Muslim-majority countries. Porch's assessment of COIN points to an opening for future research to investigate the potentially illiberal effects of certain kinds of military operations, even when countries are embedded within an ostensibly liberal alliance.

NATO's Democratizing Effects: Distinguishing between Overgeneralization and Change over Time

Claims about NATO's democratizing effects revolved around three interrelated but conflicting arguments. In addition to the assertion that NATO helped advance democracy, there were counterclaims that NATO had not done so during the Cold War and so was not equipped to do so after. In addition, some observers argued that the EU was in a better position than NATO to encourage democratic change. Overall, I find that the democratization literature modeled appropriate circumspection, sound research methods, clear specification of scope, and did not overgeneralize, though the factors that led to early democratization atrophied over time. Arguments about the implications of NATO's past failures to democratize members and the EU's capacity to secure democratization were overstated in some cases.

While the literature arguing that NATO had democratizing effects did not generally overstate its claims—because researchers were careful to specify the scope conditions under which the alliance could elicit democratic reform—democratic consolidation has nevertheless been difficult to achieve in some countries. Even in the narrower sphere of defense, some of the earlier gains in democratic civil–military governance have not advanced. Reviewing my own work, I could have explored a greater range of alternative explanations, been more skeptical about how enduring NATO's liberalizing effects would be, and shown greater awareness of how quickly and violently international politics could change.

Early literature on NATO enlargement in the post–Cold War period specified social processes and incentives that either: 1) maintained countries in the democratic fold, or 2) moved them there because of the cultivation of shared values and interests. One scholar argued that continual exposure in NATO to liberal democracies, consensus-based decision-making, and compromise constituted "self-healing tendencies" in the alliance, grounded in democratic statecraft (Thies 2009: 19–20). Others found that NATO variously shamed, socialized, or incentivized postcommunist elites into democratic compliance (Wallander 2000; Moore 2002; Epstein 2005, 2008; Gheciu 2005a, 2005b; Lucarelli 2005; Thies, Hellmuth, and Millen 2006).

In 2005, I argued that NATO had "contributed to democratization," and in 2008 that "NATO pushed all states that it engaged closer to a model of diffuse democratic civil–military relations than would probably have happened in the absence of the alliance's enlargement policy" (2008: 107–108). These claims were based on extensive fieldwork that showed differing attitudes of policymakers before and after encounters with the alliance, as well as different policies that reflected a growing preference for democratic oversight of the armed forces. I documented the curtailment of military personnel's role in politics; the increasing authority of civilians in defense planning; and growing access for parliaments, publics, and civil society organizations to defense information, including around defense budgets.

Similarly, other scholars had pointed to NATO's aspirations to create "an expanding pacific federation, informed by a common commitment to democratic principles" (Moore 2002: 25) or the alliance "engaging in political implementation of shared values and norms of liberal-democracy that have always been recognized as crucial for the existence of the transatlantic security community" (Lucarelli 2005: 85). These statements were accurate from the point in time in which they were written. But this research still might have paid more attention to the unevenness in the democratic credentials of NATO's earlier and newest members as well as the lack of instruments in NATO to confront democratic erosion, should it occur (Wallander in 2002 and 2018 actually did do this). Evidence of such variation was available at the time but was not always taken into consideration.

Gheciu's comprehensive work on NATO enlargement also did not overgeneralize. She carefully specified NATO's substantial influence as she was able to document it at the time: "In teaching new liberal-democratic norms, NATO exercised significant power," including in shaping "its socializees' interpretations of the world and ideas about proper ways of acting in the world." Interactions with NATO affected target actors' "conceptions of liberal-democratic identity and interests" (Gheciu 2005b: 973, 1009). Not only did the alliance target elites, but also public opinion and younger civilians in the defense field as well as military personnel to try to ensure that future generations would internalize NATO's definition of democratic military oversight, privileging accountability, transparency, and inclusion (see also Waterman and Zagorcheva 2001). Even if nationalist, socialist, and communist elements remained suspicious of NATO and US imperialism, the implication of these studies was that swaths of the public and the policymaking elite had accepted NATO's agenda. Many in the postcommunist world aspired to an idealized version of Western liberalism as a form of governance, however hypocritical Western liberalism actually was (see Gheciu 2019).

More recent work has called into question these earlier studies, though they do not disconfirm earlier findings or point to overgeneralization. Poast and Chinchilla, for example, examining the second round of NATO enlargement in 2004 in which seven postcommunist countries were admitted, found that "NATO membership played little role in the democratic development in Eastern Europe and the post-Soviet space" (2020: 485–486).[14] Adding nuance to this conclusion, however, they do not dispute that NATO could have enhanced democratic governance by insisting on critical institutional and personnel changes in support of civilian oversight (475–476). They also concede that NATO could have created a more secure environment in which democratic development was allowed to evolve (including particularly in the Baltic States [477]). And they acknowledge that NATO could have had a role in socializing elites (and possibly others) in democratic principles concerning military-security governance (474). But in comparing countries with a range of membership perspectives vis-à-vis the EU and NATO, Poast and Chinchilla's statistical analysis shows no clear relationship between striving for NATO membership and democratic quality when measured according to the Liberal Democracy Index from the Varieties of Democracy project, which accounts for constraints on the executive, judicial independence, strength of the rule of law, civil liberties, and minority rights, among other factors.

The counterargument is that the Liberal Democracy Index does not capture what scholars of NATO enlargement and democratization were tracking: namely, the democratization of civil–military relations. Parliamentary committees charged with military oversight, civilianized ministries of defense, and public scrutiny of the armed forces through a free and effective media are distinct from most measures of executive authority, for example. However, in contrast to some of the more fine-grained assessments of NATO's effects from over a decade ago, some recent research shows NATO has had only limited enduring impact, even in the narrower defense governance sphere, pointing to change over time, though not overgeneralization.

Whereas inculcating East European civilians with the idea that they *should* exercise significant military oversight and acquire deep military expertise *outside* of military service was an early undertaking by NATO (Epstein 2008), those ideas had not taken firm root by 2020. In fact, military security, planning, and operations expertise was closely held by active and retired military personnel who remained the go-to authorities for the most critical decision-making and guidance, side-lining civilians (Pion-Berlin, Acácio, and Ivey 2019). Postcommunist countries had become accustomed to appointing civilian ministers of defense, and there were increased opportunities for civilians to specialize in "strategic studies." But these scholars also found that military financial information was often withheld from civilians, undermining civilian capacity to participate in the most critical planning tasks (1077).

[14] In 2004, NATO admitted Estonia, Latvia, Lithuania, Slovakia, Romania, Bulgaria, and Slovenia.

With the benefit of hindsight, no one should have inferred that NATO's early post–Cold War commitment to democratization in its newest members would necessarily endure. NATO's targeted interventions, the Partnership for Peace, the Membership Action Plans, the Parliamentary Assembly and numerous other innovations and trainings were geared toward democratization. In the face of numerous wars, however, these initiatives lost momentum. But also misguided was the claim that because NATO had not actively pursued democratization in its members during the Cold War (Portugal, Turkey) it was not poised to do so after (Mandelbaum 1995; Reiter 2001). The alliance's aims did shift in ways that International Relations theory can explain, including toward a democratizing agenda in the early 1990s (Gheciu 2005a). Henrik Larsen, for example, assesses NATO's rising and diminishing interest in democratic pursuits over time by examining NATO's core powers' changing strategic and political priorities, which are in turn a reflection of changing regional and global security conditions (Larsen 2019).

The argument that democratization was better left to the EU was another form of overgeneralization with which the larger field of EU studies and policy was also grappling by the 2010s. Over decades, the EU (or its precursors) admitted a slew of new and uncertain democracies, some of which had in their very recent pasts been among the world's most brutal totalitarian regimes (Germany and Italy were, after all, founding members of the European Economic Community, established by the Treaty of Rome in 1957). Portugal, Greece, and Spain all subsequently joined the EU after emerging from different forms of dictatorship and military rule in the decades following World War II. Their murderous, divisive, and oppressive histories notwithstanding, their democratic resilience within the EU endured. None had ruling coalitions that exhibited the disdain for human rights or enacted democratic incursions on a par with Hungary, Poland, or the Czech Republic in the 2010s, which suggests the EU has less causal force than domestic politics. Maybe the EU should have developed more explicit expulsion or sanctioning mechanisms for democratic abrogation (the latter of which does exist in the EU but is difficult to use because of the consensus-minus-one provision and other political constraints).[15] But earlier episodes of democratic consolidation under less-than-optimal conditions appeared to obviate the need. This led some scholars to erroneously conclude that the EU would suffice for ECE's ongoing democratization as well. In retrospect, skeptics of NATO enlargement imputed too much democratizing power to the EU.

[15] On the EU's difficulties coercing democratic compliance from its member states, see Sedelmeier 2014.

Conclusion

Having examined the debate over NATO enlargement, I found instances of overgeneralization, unsubstantiated or inapt historical comparisons, and insertion of unacknowledged political convictions, committed by both researchers and policymakers. Strategies for avoiding overgeneralization by researchers include specifying study limits, exploring alternative explanations, and considering possible upending scenarios—practices that many scholars are encouraged to develop in their PhD programs. But in addition, and perhaps at odds with what researchers are taught, avoiding overgeneralization requires being forthright about biases, including those embedded in theory. Being clear about the evidentiary basis of one's claims as well as areas of uncertainty is essential. Finally, scholars and policymakers alike should periodically review their own work and past claims to evaluate their contributions, acknowledging when earlier findings were overstated or disconfirmed and why.

Speaking only for myself, and in light of a careful review of the literature on NATO enlargement, including my own contributions, I could have specified the limits of my studies more carefully, even if my claims about NATO's democratizing and denationalizing effects were confirmed at the time that I wrote them. I could have also expanded the range of alternative explanations I explored, along with possible upending scenarios. In addition, acknowledging the ways in which my views on the importance of democratic sovereignty evolved in the context of my research would have allowed me to point out when I and others made claims that were essentially unknowable—including, most crucially, about Russia's political trajectory. While I am not opposed to expert speculation, transparency about what one can and cannot know is essential (see DeMartino, Chapter 6, this volume). And while I suspected in 2024 that NATO enlargement had not caused Russia's invasion of Ukraine in 2022 (but rather that Russian aggression had ultimately spurred more countries to join), I also concede that Adam Tooze was correct to point out that we will only know more on this topic if Russia's archives on the subject ever become available.[16]

On many occasions, critics of NATO enlargement overstated their case. There is no evidence, for example, that the EU is a better guardian of democratic practice than many other international organizations (Epstein and

[16] See Adam Tooze, "John Mearsheimer and the Dark Origins of Realism," *The New Statesman*, March 8, 2022. https://www.newstatesman.com/ideas/2022/03/john-mearsheimer-and-the-dark-origins-of-realism.

Jacoby 2014). Some skeptics of enlargement also argued that Russia was unlikely to develop the mindset or the capability to pose a threat to its neighbors or beyond, making enlargement unnecessary. Others argued that NATO enlargement would cause a Russian nationalist revival—when in fact many variables could contribute to such an outcome. On occasion, this literature mischaracterized alternative approaches or made false claims about the origins of policies. On both sides of the debate, there was a lack of appreciation for the dynamism intrinsic to international affairs, leading to an underestimation of the ways in which institutions, threats, and actors change over time—possibly in contrast to prior research findings. Greater circumspection and more critical reflection on one's own work could help researchers provide sounder policy advice. Policymakers, meanwhile, tend not to acknowledge evidence or prospective scenarios that do not support their chosen policy, and may actively seek to limit debate.

I also found that value judgments infiltrate literatures without scholars being in direct dialogue about how central those values can be to their disagreements over findings. This is an especially thorny issue because, as researchers, many of us are trained to strive for objectivity. It is unclear, however, whether objectivity is possible, or the conditions under which it is even desirable. While ideally research can and should change our minds and inform our priors because our values evolve as we learn, many values embedded in International Relations theorizing also reinforce an inequitable status quo. Being aware and forthright about how those biases affect the claims we make—in connection with evidence but also in addition to it—would clarify what is actually being debated in the context of policy engagement.

References

Adler, Emanuel (2008) "The Spread of Security Communities: Communities of Practice, Self-Restraint, and NATO's Post-Cold War Transformation." *European Journal of International Relations* 14(2): 195–230.

Albright, Madeleine (2012) *Prague Winter: A Personal Story of Remembrance and War*. New York: Harper.

Avant, Deborah (2005) *The Market for Force: The Consequences of Privatizing Security*. Cambridge/New York: Cambridge University Press.

Barma, Naazneen and James Goldgeier (2022) "How Not to Bridge the Gap in International Relations." *International Affairs* 98(5): 1763–1781.

Bondarev, Boris (2022) "The Sources of Russian Misconduct: A Diplomat Defects from the Kremlin." *Foreign Affairs* 101(6): 36–55.

Bunce, Valerie J. (1985) "The Empire Strikes Back: The Evolution of the Eastern Bloc from a Soviet Asset to a Soviet Liability." *International Organization* 39(1): 1–46.

Carr, Edward Hallett (1945) *The Twenty Years' Crisis: 1919–1939*. London: Macmillan.

Cohen, Stephen F. (2001) *Failed Crusade: America and the Tragedy of Post-Communist Russia*. London and New York: W.W. Norton & Co.

Epstein, Rachel A. (2005) "NATO Enlargement and the Spread of Democracy: Evidence and Expectations." *Security Studies* 14(1): 63–105.

Epstein, Rachel A. (2006) "When Legacies Meet Policies: NATO and the Refashioning of Polish Military Tradition." *East European Politics and Societies* 20(2): 254–285.

Epstein, Rachel A. (2008) *In Pursuit of Liberalism: International Institutions in Postcommunist Europe*. Baltimore, MD: Johns Hopkins University Press.

Epstein, Rachel A. and Wade Jacoby (2014) "Eastern Enlargement Ten Years On: Transcending the East-West Divide?" *Journal of Common Market Studies* 52(1): 1–16.

Flockhart, Trine (2004) "'Masters and Novices': Socialization and Social Learning through the NATO Parliamentary Assembly." *International Relations* 18(3): 361–380.

Friedman, Milton (1953) "The Methodology of Positive Economics." In Friedman, *Essays in Positive Economics*. Chicago: University of Chicago Press, pp. 3–43.

Gaddis, John Lewis (1998) "History, Grand Strategy and NATO Enlargement." *Survival* 40(1): 145–151.

Gani, Jasmine K. and Jenna Marshall (2022) "The impact of colonialism on policy and knowledge production in International Relations." *International Affairs* 98(1): 5–22.

Gheciu, Alexandra (2005a) *NATO in the "New Europe": The Politics of International Socialization after the Cold War*. Stanford, CA: Stanford University Press.

Gheciu, Alexandra (2005b) "Security Institutions as Agents of Socialization? NATO and the 'New Europe.'" *International Organization* 59(4): 973–1012.

Gheciu, Alexandra (2019) "NATO, Liberal Internationalism, and the Politics of Imagining the Western Security Community." *International Journal* 74(1): 32–46.

Goldgeier, James (1999) *Not Whether but When: The Decision to Enlarge NATO*. Washington, DC: Brookings Institution Press.

Goldgeier, James and Joshua R. Itzkowitz Shifrinson (2020) "Evaluating NATO Enlargement: Scholarly Debates, Policy Implications and Roads Not Taken." *International Politics* (57): 291–321.

Grigas, Agnia (2016) *Beyond Crimea: The New Russian Empire*. New Haven: Yale University Press.

Jacoby, Wade (2004) *The Enlargement of the European Union and NATO: Ordering from the Menu in Central Europe*. New York: Cambridge University Press.

Kennan, George F. (1997) "A Fateful Error." *New York Times*, 5 February: A23.

Lanoszka, Alexander (2020) "Thank Goodness for NATO Enlargement." *International Politics* 57: 451–470.

Larsen, Henrik B. L. (2019) *NATO's Democratic Retrenchment: Hegemony after the Return of History*. London/New York: Routledge.

Lucarelli, Sonia (2005) "NATO and the European System of Liberal-democratic Security Communities." In *Socializing Democratic Norms*, ed. Trine Flockhart. London: Palgrave Macmillan, pp. 85–105.

Mandelbaum, Michael (1995) "Preserving the New Peace: The Case Against NATO Expansion." *Foreign Affairs* 74(3): 9–13.

Mearsheimer, John J. (2014) "Why the Ukraine Crisis is the West's Fault." *Foreign Affairs* 93(September/October): 1–12.

Mearsheimer, John J. (2018) *The Great Delusion: Liberal Dreams and International Realities*. New Haven: Yale University Press.

Mearsheimer, John J. (2022) "John Mearsheimer on the Causes and Consequences of the Ukraine War." June 16, European University Institute. https://www.youtube.com/watch?v=qciVozNtCDM.

Moore, Rebecca R. (2002) "NATO's Mission for the New Millennium: A Value-Based Approach to Building Security." *Contemporary Security Policy* 23(1): 1–34.

Pielke, Roger A. Jr. (2007) *The Honest Broker: Making Sense of Science in Policy and Politics*. Cambridge: Cambridge University Press.

Pion-Berlin, David, Igor Acácio, and Andrew Ivey (2019) "Democratically Consolidated, Externally Threatened, and NATO Aligned: Finding Unexpected Deficiencies in Civilian Control." *Democratization* 26(6): 1070–1087.

Poast, Paul and Alexandra Chinchilla (2020) "Good for Democracy? Evidence from the 2004 NATO Expansion." *International Politics* 57: 471–490.

Porch, Douglas (2013) *Counterinsurgency: Exposing the Myths of the New Way of War*. Cambridge/New York: Cambridge University Press.

Reeves, Christopher (2019) "From Intervention to Retrenchment: Poland's Strategic Culture and the 2011 Libyan Campaign." *Europe-Asia Studies* 71(7): 1140–1161.

Reiter, Dan (2001) "Why NATO Enlargement Does Not Spread Democracy." *International Security* 25(4): 41–67.

Rothschild, Joseph and Nancy M. Wingfield (2000) *Return to Diversity: A Political History of East Central Europe since World War II*. New York: Oxford University Press.

Russet, Bruce and Alan C. Stam (1998) "Courting Disaster: An Expanded NATO vs. Russia and China." *Political Science Quarterly* 113(3): 361–382.

Sarotte, Mary Elise (2019) "How to Enlarge NATO: The Debate inside the Clinton Administration." *International Security* 44(1): 7–41.

Sarotte, Mary Elise (2021) *Not One Inch: America, Russia, and the Making of Post-Cold War Stalemate*. New Haven: Yale University Press.

Satter, David (2016) *The Less You Know, the Better You Sleep: Russia's Road to Terror and Dictatorship under Yeltsin*. New Haven: Yale University Press.

Schimmelfennig, Frank (2003) *The EU, NATO and the Integration of Europe*. Cambridge: Cambridge University Press.

Sedelmeier, Ulrich (2014) "Anchoring Democracy from Above? The European Union and Democratic Backsliding in Hungary and Romania after Accession." *Journal of Common Market Studies* 52(1): 105–121.

Stone, Deborah (2018) "The 2017 James Madison Award Lecture: The Ethics of Counting." *P.S.: Political Science and Politics* 51(1): 7–16.

Thies, Wallace J. (2009) *Why NATO Endures*. New York: Cambridge University Press.

Thies, Wallace J., Dorle Hellmuth, and Ray Millen (2006) "Does NATO Enlargement Spread Democracy? Evidence from Three Cases." *Democracy and Security* 2(2): 201–230.

Vachudova, Milada Anna (2005) *Europe Undivided: Democracy, Leverage, and Integration After Communism*. Oxford/New York: Oxford University Press.

von Hlatky, Stéfanie and Michael Fortmann (2020) "NATO Enlargement and the Failure of the Cooperative Security Mindset." *International Politics* 57: 554–572.

Wallander, Celeste A. (2000) "Institutional Assets and Adaptability: NATO after the Cold War." *International Organization* 54(4): 705–735.

Wallander, Celeste A. (2002) "NATO's Price: Shape Up or Ship Out." *Foreign Affairs* 81(6): 2–8.

Wallander, Celeste A. (2018) "NATO's Enemies Within." *Foreign Affairs* 97(4): 70–81.

Waterman, Harvey and Dessie Zagorcheva (2001) "Correspondence: NATO and Democracy." *International Security* 26(3): 221–235.

Williams, Margit Bessenyey (2002) "European Integration and Minority Rights: The Case of Hungary and its Neighbors." In *Norms and Nannies: The Impact of International Organizations on Central and East European States*, ed. Ronald H. Linden. Lanham, MD: Rowman and Littlefield, pp. 227–258.

Zvobgo, Kelebogile and Meredith Loken (2020) "Why Race Matters in International Relations." *Foreign Policy*. June 19. https://foreignpolicy.com/2020/06/19/why-race-matters-international-relations-ir/.

8
Data Ethics and the Power of Data

César Caballero Reinoso

Responsibly using research in policymaking can sometimes require balancing personal and professional concerns with those related to politics, ethics, and data quality.[1] In this chapter, I draw on some of the concepts and dilemmas developed in other chapters of the volume to articulate a framework to balance some of these trade-offs. This approach takes into account the professional and personal risks of contesting unsound uses of research—such as losing status or facing other forms of backlash—relative to the benefits of retaining one's peace of mind and ethical center by advocating for evidence-based policy. However, going along with possible unsound uses of research may also hold benefits, such as of retaining one's position, office, or influence, which can allow one to shape how information and research are used and agendas are set in the future. Yet, in the short term, this can involve the cost of producing suboptimal policy outputs. It can also lead to a kind of moral injury, such as compromising one's professional position or reputation by being used for "box-checking"—providing political cover for an already-decided policy decision or stance.

I illustrate the application of this framework based on examples from recent contentious policy debates in Colombia, as well as my personal experiences as a researcher in government and in the private sector. Most centrally, it is helpful to anticipate potential policy engagement risks to be able to make sound decisions when they arise—something I could have done better in several instances.

A first example illustrates how one may not always be prepared to interact within the "black box" of the policymaking process. In the mid-1990s, I worked in Colombia's National Planning Department (DNP) and the Tax Revenue Commission as a technical advisor. In one of my first meetings, an official was discussing the national budget deficit. He showed me a table,

[1] This account is based on an interview with Oliver Kaplan that was transcribed by Michelle Carrere-Seizer and translated to English by Oliver Kaplan. Mary Cates, Danielle Gilbert, and Paula Pinzón provided valuable comments.

César Caballero Reinoso, *Data Ethics and the Power of Data.* In: *Speaking Science to Power.* Edited by: Rachel A. Epstein and Oliver Kaplan, Oxford University Press. © César Caballero Reinoso (2024). DOI: 10.1093/oso/9780198875185.003.0008

repeatedly using the expression "Above the line, below the line." I looked at the table once. I looked at again. It didn't have any lines. I asked another official beside me what I was missing. He pointed to a little blank space between two rows of data. He explained that a line had been drawn there before, but that the economists decided to erase it. They had developed a technical language, one that makes them look like experts, as if perhaps they don't want others to understand them. It turns out the invisible "line" in the summary table distinguished between current and noncurrent income in the statement of fiscal accounts. And, indeed, only those with technical training and experience in fiscal and budget issues would understand the meaning of that shorthand.

As this story illustrates, data, in general terms, and statistics, in particular, have a function of legitimizing the discourses of power. For many political, social, and business leaders, narratives matter. Such individuals often use data to promote the understanding of complex issues or as a mechanism to increase the credibility or legitimacy of their decisions. Yet they also sometimes use data to *confuse*. If you manage to sound credible and people do not understand, most of them will be too embarrassed to ask a question. That is where a first barrier arises in making sense of numbers.

Data has enormous power to communicate and create realities, but it can also obstruct open dialogues from occurring. One of the social functions of statistics and data is to build bridges of communication. In a discussion, even a difficult one, when the different parties agree to engage with the same information there is a starting point for understanding and construction of meaning.

In this chapter, I explore both responsible and questionable uses of data in policymaking. I draw on examples and insights from my career as both a practitioner and a generator of data from my time as the director of a census bureau and later as a pollster. Along the way, I encountered many of the responsible engagement issues identified by the scholarly contributors to this volume: transparency about sources and methods and communicating uncertainty, coping with unpopular findings, the risks of seduction by power, and setting clear moral standards and redlines. I also participated in partnered engagement when working with my company's employees to collectively determine our ethical north stars related to the social issues we held dear. In the end, scholars should seek to stand their ground in the face of political pressure to manipulate evidence, even if it sometimes means weighing the risks of making professional and perhaps personal sacrifices.

Research and the Colombian Census Bureau

As director of the Colombian National Census Bureau (the Departamento Administrativo Nacional de Estadística, or DANE by its Spanish initials) from 2002–2004 at the beginning of Álvaro Uribe's presidency, I strove be an effective spokesman when it came to data, and to humanize the figures. At the start of the administration, the government accepted all the figures DANE put out—even the unflattering ones—as they were evaluating the previous government. So, while the data showed a critical state of affairs in the country, the data served useful purposes at that time, including to feed the discourse of decision-makers, criticize the opposition, and disparage the efforts of the previous government. And the new president criticized his predecessor, Andrés Pastrana, not only because of the failed peace process with the Revolutionary Armed Forces of Colombia (FARC), but also because of his closeness to Fabio Valencia, Uribe's political opponent in his home department of Antioquia.

When I was appointed as the director of DANE, I remember making two requests, which I now recognize as sorts of "redlines." First, I wanted to be able to appoint my deputy director; I did not want them to be imposed on me as a political decision. Second, I knew that at some point DANE would release data that could be unpopular. I vowed to always release it, no matter the possible controversy or professional risks.

I first met Álvaro Uribe in December 1998 when I was studying for my Master's degree at Oxford University. He came to Oxford having just finished his term as governor of Antioquia, a department (state) in Colombia's northwest region. We developed a good relationship, and when I returned to Colombia he put me in charge of drafting his political platform. In the context of a political campaign, the use of data is important for developing campaign strategies, and it is also perhaps a calmer moment than when in office, with fewer political and policy challenges. But, at one point during Uribe's presidential campaign, when I was his policy director, a challenging event occurred.

It was May 1st, Labor Day in Colombia, and a big event was planned to showcase Uribe's proposals on worker issues. Campaign policy experts put together a document that collected the candidate's proposals on the topic of employee benefits—one of Uribe's core concerns that was well-received by the business community. He argued that reducing worker benefits would reduce the unemployment rate. However, the available studies and evidence did not support that statement. I proposed including a paragraph pointing out the risk of this kind of policy. He also asserted that the reduction in overtime and Sunday hours would generate a million additional jobs during nights and weekends. But there were also no studies or evidence that supported this claim, either, and I was not inclined to include it. Because of this disagreement, I soon resigned as his policy advisor. Yet, a few months later, I was asked to return to the campaign.

Since Uribe was an opposition candidate up against a highly unpopular incumbent government that was doing badly by almost all indicators, criticizing them with their official figures was easy. So, once he won the election, the credibility of the data from the Pastrana period and the methods with which they were collected was beyond question. After all, the DANE figures, no matter how bad, were referring to the management of the previous government, and that fed the new government's narrative. So, everyone was happy with the numbers, and that was that. During the first months of the government, things went well, and I felt supported by the president and his economic team.

A dilemma arose, however, when data appeared some months later that showed that the new government was not performing well in some areas. From my point of view, the president was forthright about the indicators, and when results were positive, he mentioned them in his speeches. But the largely poor figures at that time posed a problem, and DANE faced some questions.

Early Concerns: The Constitutional Anti-Corruption Referendum

I had some early concerns about the use of government-generated data in relation to the 2003 Constitutional Anti-Corruption Referendum (Aznárez 2003). Based on input from DANE analysts, it became clear that there was a delay in the Colombian electoral census administered by the National Registrar's Office that was required to hold the referendum. By contrast, DANE had more accurate information about the electorate, and could produce it more rapidly. This first episode highlights the links between data collection and management and electoral outcomes.

The confirmation of the voter rolls in Colombia runs as follows in cases when voters may have died. When a person dies in Colombia, a "death certificate" is issued in triplicate: one stays at the clinic or place where the death is declared, another is sent to DANE, and the third is delivered to the relatives (so that they can legally register the death). However, the process is impeded for three types of people: those who live in isolated areas or in such poverty that they do not report the case; victims of forced disappearance during the armed conflict (who may still be alive, and therefore must remain on the electoral rolls); and wealthy individuals who, to avoid paying inheritance taxes, continue to manage the assets of the deceased and have an incentive not to register their relative's death.

Together with the Registrar, we devised a process in which that office would pass the electoral census to DANE, and then DANE would update the registry (DANE has a requirement of statistical secrecy, but the Registrar does not, so the Registrar could legally run the census). Budgetary support was required to complete the process, so we jointly sent a request letter to the president, and he delegated the issue to the Minister of Interior. The polls showed that the referendum would win by a large margin, and therefore the ministry was not inclined to allocate the necessary resources. The initiative to update the electoral census ground to a halt.

The day the referendum was lost, the government was not happy with the results, and I was summoned to a meeting at the Presidential Palace, the Casa de Nariño. The problems with the electoral census were apparent, and the government had also known of the possible extra voters on the rolls—over-registrations in the census—more than a year in advance. The Registrar and I had identified a way to fix the issue and had pleaded that it should be done prior to the election. I felt the Registrar should be supported even in the face of the election loss.

Talking to several presidential aides with my back to the door, I had not realized that the president had just arrived. I turned around to see the president looking at me. I reaffirmed my comment to him: "Mr. President, the referendum was lost, and it is democratic to accept it when that happens." With that, I left. The loss of the referendum was not the desired outcome, and it perhaps could have been avoided or at least foreseen with better information about the electorate. But the result had to be accepted nonetheless.

With more time and distance, I realized the corruption referendum episode was not an isolated incident and maybe did not fit with my vision of public service.

An Unpopular Assessment of Colombia's Highway Security Program

In January 2003, a DANE analyst walked into my office. There was a datapoint that did not make sense to him about the highway security program "Long Live Colombia, and Travel Through It" (*Vive Colombia, Viaja Por Ella*). The news media had widely reported on the "success" of the program. It had been put in place so that Colombians could travel safely by land after years of facing risks of insurgent groups and kidnapping. In this period, the FARC came to commonly implement a practice known as "miracle fishing" (*pesca milagrosa*), whereby they would set up roadblocks and kidnap people from their cars to either extort money or kill them (Rubio 2005). In retrospect, this would become a key instance of grappling with "unpopular findings" (Olsen, this volume).

The analyst reported that he did not know where the encouraging figure the president announced of 20 million people traveling by road during a two-week period had come from. If it were accurate, it should have been observable in household spending indices, and therefore in inflation. But, based on the available data, the numbers did not add up. A detailed investigation of the result contrasted it with other sources. The official vehicle registry showed that there were only 3.3 million vehicles in the country in 2003—a far cry from the 20 million motorists that had been put forward.

In the following cabinet meeting, I commented that DANE had some doubts and explained the impossibility of the figures generated by the Ministry of Transportation. The president requested a meeting with the Minister of Transportation to clarify the matter. Yet Easter came, and the government again spoke of the success of the program. But this time, it said that 10 million people had traveled in 2.6 million vehicles on the roads. The same exchange was repeated in the next cabinet meeting, and I pointed out issues with the data. In June, during the school holidays, they announced success once more, with a figure of 8 million travelers. I again showed the inconsistency between what was said and what was possible in reality.

When the meeting with the Ministry of Transportation finally happened, a police lieutenant arrived and explained how the information was calculated. He had asked for all the data on how many cars went through all the tollbooths. He multiplied that by four people per car. But he had not taken into account that a couple on vacation can go through ten tolls one way and another ten on the way back. That would be only two people, but he would count them as 80 tolls paid, as if there were four people in the car! When lower-level bureaucrats like the police officer meddle with statistics that they do not understand, issues can arise. Though doing his best in the absence of specific directions, and not meaning any harm, his inventiveness with the data almost led to a costly instance of uncorrected misinformation. As a result of the exchange, the government began to bring down the numbers, until, in the end, the government's balance for the program gave a figure of 4 million people.

As government administrations proceed, they sometimes use statistics to highlight their achievements but may develop attitudes of denial in the face of figures that do not suit them. If someone tries to point out that things are not working perfectly or the results are not that great, they become the odd one out. That eventually happened to me, even while I viewed one of my roles as being a containment dam—to protect the census agency and its institutional tradition and integrity.

Slippery Slopes and Massaging the Victimization Survey

I felt that DANE could and should involve itself in subjects beyond traditional economic issues. Indeed, social issues are also included in its legal mandate. In 2003 DANE published its first analysis with a gender perspective, its first survey of political culture, and an institutional performance survey. DANE also carried out the first Victimization Survey in 2004, an initiative of the World Bank and National Planning Department (DNP; El Tiempo 2004b). The case of the Victimization Survey would raise issues of possible ethical "unpopular findings" (Olsen, this volume), "slippery slopes" (Lyall, this volume), and putting a positive face on data results (DeMartino, this volume).

To expand the sample size of the survey, resources were requested from the mayors of Bogotá, Cali, Medellín, and Barranquilla, whose cities were being surveyed. An early analysis of the results suggested that they could be problematic for the administration. In his campaign, Uribe attacked the Pastrana administration's creation of the demilitarized zone for negotiations with the FARC. He argued that because of Pastrana's decision criminality had been allowed to run rampant.

The Victimization Survey revealed two complex issues for the government. First, it showed that residents of Medellín were reporting that the demobilized members of the Cacique Nutibara Bloc of the AUC (Self-Defense Forces of Colombia) paramilitaries were extorting people. Second, the police data showed that during the weekends when the highway security program "Long Live Colombia, and Travel Through It" was implemented, cases of common crime increased in the cities. It was only natural that if the president announced that the police and the army would be monitoring the highways over the weekend, the manpower available to secure the cities would decrease and criminals would take advantage of this.

I sent a summary of these issues to the president, saying that I would soon present it to members of the cabinet. I had meetings with the Defense Minister, the director of the intelligence agency known as the Administrative Security Department (formerly DAS), the director of the national police, and the president's communications adviser. I was also going to present the results to the mayors of Colombia's four largest cities. I never received any complaints about the results, my presentation of them, or the technical quality of the study.

On the evening before a press conference on this data, I received a call from one of President Uribe's main advisers. He stated that the government did not agree with the data, and that for that reason he was communicating to me to either change them or not publish the survey.

I refused the order. I said that the data would not be altered and that the survey would be published. From that moment until 10 p.m. that night, I tried to reach the president by phone. I spoke with several different officials, and contradictions were evident in all of these conversations: they expressed their support for my position, but suggested that following the order was the best option for me to continue as director of DANE.

My main concern was the Population Census slated for 2005. It was the key task of my position, and it was already late. The government did not want to carry it out for fiscal reasons, but thanks to a mobilization effort involving mayors, governors, academia, unions, and social organizations, a joint financing scheme was put in place to make it viable. I believed that my departure from the agency would cause problems for the census.

Several of my confidants suggested options such as postponing the presentation of the results for a few days or publishing the data without a press conference. But I believed that these alternatives would only do more harm to the agency, to the official statistics, and to the government. I now recognize this episode in terms of a possible case of sharing certain messages to maintain popular support for the government and reassure Colombian citizens (similar to the issues referenced in DeMartino's chapter [6]), but this could come at the larger cost of damaging the credibility of the government in terms of how it manages information.

When government officials do not stick to their core principles, this can pave the way for increasingly problematic requests (similar to the ethical "slippery slope" discussed by Lyall in Chapter 2). After giving in to a first inappropriate instruction, it would all go downhill. For this reason, I wrote a brief note in which I expressed my decision to resign: "Much to my regret, I received an order that I feel morally incapable of carrying out" (El Tiempo 2004a). My resignation made national news and was featured on the cover of Colombia's main weekly magazine, *Semana*, with the headline "He Who is a Gentleman Resigns"—a double-meaning play on my last name (Semana 2004; Figure 8.1).

What followed is a matter public record. Several rumors circulated about me, suggesting that I had engaged in professional misconduct: that I had

Figure 8.1 In 2004, César Caballero's exit as director of the Colombian census bureau (DANE) was featured on the cover of Semana magazine (Colombia) with the headline "He Who is a Gentleman Resigns," a double-meaning play on his last name.
Semana

refused to hand over information to other cabinet members in advance of meetings, that I would soon be prosecuted by the Attorney General's office, and that I was involved in corruption scandals. I also found that my employment options had become limited, and that certain job prospects disappeared. I even faced threats to my life.

This episode reflects the pressures that researchers can encounter when dealing with unpopular findings, and the trade-offs and balancing decisions that sometimes must be made (e.g., Kaplan 2021). Figures are a device of power, which can be used positively or negatively, to build or to destroy. Societies need transparency, rigor, and credibility in official statistics. I made decisions to ensure data would be used only positively, and I encountered personal risks and costs based on those decisions. At the same time, certain doors were opened for me in the areas of international cooperation, academia, and the private sector. And having a clear conscience is priceless.

Partnered Engagement and the Ethics of Conducting Polls

Many hold the supposition that pollsters, as social researchers, can and "should" be neutral. In other words, it is a commonly held belief that there are no biases in surveys—not at the moment that they go out to the field, nor in the choice of data collection methods, nor in the presentation of results. These are seen as the product of a totally neutral process.

Yet data collection does not always work like this. As a social researcher, one can be transparent and rigorous and work to correct biases, but I do not agree with being "neutral." Why? Because any choice related to these elements, from the type of project that is implemented to who finances it, mediates between researchers and their results (e.g., Herbst 1993).

Everybody has some biases, and when it comes to polling, people should be clearly aware of pollsters' positions—similar to what Epstein (Chapter 7 in this volume) advocates with respect to researcher transparency about their political priors. In other words, they should not be blinded by the veil of "supposed" objectivity. For this reason, at my company, Cifras & Conceptos (Data & Concepts), a partnered engagement process (Kaplan, this volume) brought about a policy of clearly and explicitly stating the social causes in which we believe on our web site and social media outlets. In this sense, transparency can be weighed as more valuable than a false pretense of neutrality.

The firm has carried out many studies on the views of Colombian society on issues such as peace processes, abortion, the legalization of marijuana, and feminism, among many other polarizing topics. In all these cases, the firm publicly expressed its support for these causes. Even when the data show that the vast majority of Colombians do not support an issue, the firm has always published the figures anyway, to ensure the transparency and rigor of the data. Publishing results even when they are contrary to our personal beliefs lends credibility to those results, as well as the general reputation of our future polling projects.

This presents a clear ethical dilemma: Can or should a pollster state their own positions on key issues? By virtue of owning a public opinion firm, should a pollster give up their right to holding or expressing opinions on issues on the public agenda? Some may answer that pollsters should not share their opinions or (ideological) commitments, but there are times when it makes sense to do so.

A specific instance from 2017, at the start of 2018 presidential campaign, illustrates setting aside one's commitments to ensure data transparency. There were many candidates in the running, people who believed they had a chance of being president. But the political landscape was uncertain, so in addition to the declared candidates an open-ended question was included about who prospective voters would like to vote for as president in 2018. Normally, the most well-known names always appear along with the option of "Don't know or No response" and other lesser-known candidates, but the results indicated strong support for the political views of former President Uribe. More precisely, the survey data showed that a sizable share of voters wanted to support the former president's candidate, whoever he or she might be.

No one had ever published that kind of open-ended response option, and it would generate a lot of comments and criticism. It also reflected the former president's immense sway over voters, which had never before been so explicitly documented. What was the right thing to do? The figures had to be published to show the country that a substantial number of voters were willing to vote for "Whoever Uribe says," no matter who the candidate was. In 2018, the country elected Iván Duque as president—the candidate supported by the former president.

In some cases, candidates request to commission a survey conditional on it showing them as the winner. It may come as a surprise that they would consider it an option or have even dared to propose it—that in some cases they would make pleas for a few extra polling points. Cifras & Conceptos has never done that. Clients may contract for a study that employs a specified method, but particular results would never be promised.

It is helpful for candidates to appear in the polls, and even better if they come out in good standing. Appearing in a poll avails politicians of access to the media and funding sources, and also increases their viability among their own constituents and their ability to form political alliances. For that reason, the door to being included in a poll should not be closed to anyone, because if a candidate is not measured they are essentially rendered invisible. For instance, in the 2018 campaign, my firm assessed all sixty-four of the people who at some point expressed aspirations for the presidency. We conducted various periodic measurements and tried to treat each candidate the same. If, after a couple rounds of surveys, a candidate still has not broken through, they will be removed from the subsequent rounds. Polling also becomes simpler once the official electoral period begins, since there are fewer candidates and every officially registered candidate is included in the polls.

Cifras & Conceptos took an approach of being inclusive to avoid making anyone invisible. In addition to the emphasis on inclusion, there are several other guiding principles when conducting surveys. First, data should never be falsified and has to be published or delivered to the client in its most accurate form. Second, and relatedly, pollsters should avoid formulating biased questions. For example, in the 2022 campaign for president, a polling firm in Colombia asked respondents if they wanted to vote for a right-wing candidate, "Federico Gutiérrez"; a center-left candidate, "Sergio Fajardo"; or a leftist candidate, who was listed in the poll as "former leftist guerrilla Gustavo Petro." Clearly, presenting a candidate as a former guerrilla in a country like Colombia would do them no favors.

It will not come as a surprise that Petro did poorly in that survey, likely an effect of the wording of the question. That should be avoided. The polling industry has a voluntary international code of ethics, the ESOMAR code, and there is a set of standards that a firm must meet to receive a quality management certification (ESOMAR 2007).[2] But, in the end, ethical surveying practices primarily occur during the act of conducting a poll. Polling is both the application of a technical instrument and a form of art. In each new context, the pollster has to present balanced information.

In debates about public opinion, there are also misconceptions about the goal of political polls, which can affect whether they are used responsibly or not. The public, academics, and the media often assume that a good poll is one that predicts the future. But polls cannot predict the future. This is like having a Ferrari and complaining that it does not fly—it is a great car, but it is not designed to fly, just as surveys are a good instrument for revealing what people think today but not what they will do in the future.

The firm began to encounter some concerns about electoral polling projects. There can be significant reputational risks in publishing election poll results because, again, surveys are devices of power that are used by different actors in different ways (akin to the policy "black box" problem referenced throughout this volume). If someone finds the result of a survey useful, they usually will not ask about the method, the firm's rigor, or the quality of the data: "It works for me and says what I believe, so I will support

[2] See https://esomar.org/standards.

it and disseminate it." By contrast, if a poll result is seen as undesirable, people will attack and minimize it.

The consequences can also be more severe than the reputation of the firm. Indeed, to poll in Colombia is to poll in a society with a tradition of eliminating political opponents by killing them. The risks go far beyond economic blockades, for example. There are real risks to one's life. Perhaps when practicing the profession in safer countries, such as the United States or Switzerland, one's decisions about whether and how to study elections would probably be different and less encumbered (though such political violence is now also sadly a rising trend across Western democracies). There can be a tension between the responsibility toward one's country and generating helpful information for our electoral processes versus other obligations, such as toward my family, which affects one's calculus.

With these considerations in mind, in 2018 the company made a collective decision as an organization to follow a path that allows us to continue as a public voice and act responsibly while minimizing potential risks: We stopped asking questions about individual candidates and political races and focused on issues of public concern. The decision to not carry out electoral surveys was suggested by other members of the team, and I resisted it for several months, but finally I came to see it as the best option. This decision took us out of a segment of the market with high impact and important revenues for the company.

The firm handles ethics as a collective decision that is discussed and debated openly among our team members in a type of partnered engagement (Kaplan, Chapter 5, this volume). This eventually shapes our downstream products and public engagement. The firm has a diverse and multidisciplinary team, including a combination of more experienced and less experienced people and staff from various professional backgrounds: economists, anthropologists, statisticians, psychologists, engineers, political scientists, administrators, and even fashion designers. Dissonant voices are also encouraged to help avoid "groupthink." For example, a deputy manager of the firm had very individual views on almost all issues, from policies to management. He always expressed these freely, and sometimes his reasoning prevailed over mine. Collective reflection among various members of the firm is encouraged so that everyone's voice counts and is valued. This promotes greater balance and ensures that decisions do not simply reflect a leader's personal preferences.

In the case of withdrawing from conducting electoral polling, the deputy manager suggested that this issue created a reputational problem for the firm and would make the company more clearly part of the political debate. Over time, as the team found out about the threats made against me and that the company had needed to hire escorts to accompany me to ensure my safety, this affected their peace of mind and provoked criticism of the decisions surrounding our portfolio of projects. If an organization incurs costs for doing this type of work, and the costs do not offset the benefits of emotional or physical tranquility, then clearly the risks involved are not economically justifiable.

The decision to abandon electoral polling was not made overnight. It was a process of gradual persuasion, during which my wife became involved as we began to receive personal threats and my infant son's life was possibly in jeopardy. The combination of internal pressure from our team, the influence from my family, and my own reflection on the costs and benefits of polling all contributed to the decision. Ultimately, the firm continues to generate quality information for public debate—the true north, the guiding star—only the strategy to reach that goal has been adjusted.

Final Reflections

Whether in the context of government research or private sector polling and research, the issues surrounding responsible research and policy engagement are in many ways similar to being an academic scholar. There are some of the same dilemmas of unpopular findings, seduction and redlines, prosocial lying, and inclusion and partnerships—even though in those moments I did not have such terminology at hand to characterize those situations with precision. Yet, the nature of my practitioner positions both within and outside of government in Colombia has meant that my experiences have also differed from the realm of academia. My experiences have often been much closer to the high-stakes decisions being made by top-level politicians—politicians who were relying on data that I furnished them with. In contrast to university-based scholars, my research products also had a much clearer veneer of officialdom, since I was frequently involved with generating data from within government, or I produced public polls that would directly speak to contemporaneous political and policy issues.

Through my governmental and private sector research experiences I independently arrived at some of the same engagement guidance developed by the scholars in this volume: to articulate redlines, call out lies, and partner with colleagues. If this account offers any lessons, a central message is that it is possible to behave ethically with data and knowledge, even when facing intense political pressure and even when at great personal risk. If I was able to bear the costs of acting responsibly, then other political actors and scholars can also take the time and effort to consider how to act responsibly in their engagements. As a corollary, not doing so can reinforce the "confirmation bias" of the rulers in power. This paves the way for the development or maintenance of unsound public policies and undermines the future credibility of evidence-based policymaking.

References

Aznárez, Juan Jesús (2003) "Colombia vota contra la politquería" (Colombia Votes Against Politicking). *El Pais*, October 26, 2003.

El Tiempo (2004a) "Renunció el director del DANE" (DANE Director Resigns). September 16, 2004. https://www.eltiempo.com/archivo/documento/MAM-1509692.

El Tiempo (2004b) "Las otras cifras sobre victimización" (The Other Figures about Victimization). September 18, 2004. https://www.eltiempo.com/archivo/documento/MAM-1513776.

ESOMAR (2007) "ICC/ESOMAR International Code on Market and Social Research." December 2007.

Herbst, Susan (1993) *Numbered Voices: How Opinion Polling Has Shaped American Politics.* Chicago: Chicago University Press.

Kaplan, Oliver (2021) "Balancing Rigor and Relationships in Collaborative Research." *PS: Political Science & Politics* 54(3): 544–548.

Rubio, Mauricio (2005) *Del rapto a la pesca milagrosa: Breve historia del secuestro en Colombia.* Bogotá: Universidad Externado de Colombia.

Semana (2004) "El que es Caballero renuncia" (He who is a Gentleman Resigns). September 18, 2004. https://www.semana.com/nacion/articulo/el-caballero-renuncia/68221-3/.

9

Policymaker's Epilogue

From the Receiving End of Responsible Engagement

Leanne Erdberg Steadman

Introduction

It is no surprise that timely and compelling academic research can significantly influence a particular policy decision. But it is puzzling that sometimes research may not impact decision-making at all. Predicting where one's research and engagement will fall on the continuum from direct impact to irrelevance for policymaking is complicated, in no small measure because policy processes themselves are so dynamic. I explain this complexity so that researchers are better informed about what questions to ask their policy interlocutors before they engage. I also contend that responsibly "bridging the gap" between the research and policy worlds depends on policymakers becoming more sophisticated consumers of scholarship.

Substantial strides have been made in bridging the gap between academia and policy circles in terms of national security and foreign policy, and yet scholarship may nevertheless be ignored (Avey et al. 2022; also see the Introduction to this volume). For their part, researchers can work to better understand the policymaking ecosystem and calibrate how best to engage with it. In addition, policymakers can become more literate about academic research and its most effective uses. In this chapter, I discuss how academics can better understand policy processes and how they can become better equipped for ethical decision-making and engagement. Second, I illuminate the perils for policymakers if they fail to understand researchers' ethical dilemmas, incentives, biases, and practices and how these can impact the research they consume. By highlighting policymaking complexity, I signal to researchers how their work may be used—including in ways they may not expect. Further, I provide insight for policymakers on the possibility of misaligned incentives between researchers and practitioners, which in turn could affect how policymakers interpret and apply research.

Leanne Erdberg Steadman, *Policymaker's Epilogue*. In: *Speaking Science to Power*. Edited by: Rachel A. Epstein and Oliver Kaplan, Oxford University Press. © Leanne Erdberg Steadman (2024). DOI: 10.1093/oso/9780198875185.003.0009

A quick scan of this landscape shows that policymakers interact with academics and researchers in many ways. Some interactions are governed by statute or administrative policy, such as federal advisory committees or other appointed advisory boards. Many agencies, departments, and offices independently contact academics and researchers, commissioning research projects, organizing academic exchanges and conferences, or occurrences such as speaking invitations or attending events. Former academics, recent graduates, and current students are employed at policymaking institutions, and they bring with them personal knowledge and connections to academia. The reverse also happens, as former policymakers land in academic positions, and some are prominent public figures or public intellectuals. Think tanks and government-funded research organizations bridge research and policymaking, sometimes acting more like researchers and sometimes as shadow policymakers. Lastly, and most obviously, policymakers are members of the public and engage with journalism, widely disseminated articles, and commentaries.

Given the multiple modes of engagement, there is no single view that encapsulates what types of inputs policymakers receive from academic engagement or through what channel research has the greatest influence. For this chapter, I have selected just two aspects of policy engagement, based on my practitioner experiences, that convey clear lessons. My hypothesis is that greater understanding of how policy decisions are formed can help academic researchers tailor their own engagement. Similarly, policymakers become more informed and participatory consumers through greater understanding about research processes, including the incentives and perils that academics who pursue policy-relevant scholarship encounter.

My observations stem from having previously worked at the State Department, the National Security Council staff, and the US Institute of Peace, where I had a variety of opportunities to interact with research and researchers. These included regular exchanges and meetings with scholars, programmatic expenditures that funded research, and conferences to which academics were invited to speak. Moreover, several researchers, think tanks, and NGOs regularly sent officials (including me) their findings in reports and articles, and I regularly read publicly available articles penned by experts as a matter of general interest. Inside policymaking institutions, I also saw a growing movement toward more evidence-based decision-making, including in foreign policy and national security, of which academic research is an important component. Policy incentives to stress-test ideas with outsiders also created openings for academic engagement. As such, empirical, rigorous, and field-based findings were regularly welcomed as helpful inputs into the policymaking system. I believe the varied nature of the engagements I observed provides an opening to remark on ways to enhance understanding on each side—academic and policymaker—through the lens of ethics and responsibility.

One key takeaway for me has been that even with this engagement, structured approaches to utilizing outside research were lacking compared to other inputs to policy decision-making. On the most pressing issues, we only inconsistently asked questions such as "Who has studied this? How does the field as a whole view this topic? What are emerging areas for study? What has gone wrong in the past?" Conversely, policymakers rarely made any decision without asking more practical questions, such as "What does the intelligence say? How many resources are devoted to this? What agencies should be a part of this effort?" and even "Who are the private sector and nongovernmental actors with whom we should be working?" Policymakers often had episodic, highly specified, or ad-hoc engagement with external academics or research, calling upon such topical expertise when needed, but not necessarily seeking great breadth or depth of understanding. And we rarely considered how policy-relevant research may have been shaped by a researcher's own ethical considerations.

The chapters in this book, covering everything from the perils of prosocial lying and selection bias to overgeneralization and problems of positionality, shed light on certain ethical and responsible engagement concerns for scholars. But these topics also beg additional ethical introspection for policymakers. After all, if a scholar might be overgeneralizing to a policymaker (consciously or not) about their findings or failing to provide a platform for the local owners of the research, these problems should also figure into how policymakers choose to weigh and use the information they are consuming. With that in mind, I argue that policymakers must also consider the ethical, moral, and accountability issues associated with academic engagement.

I first describe the complexity of policymaking processes in the hopes of enhancing scholars' understanding of their possible points of entry, which should help them retain agency in the process to better assess how to responsibly engage. Second, I provide an application of this issue through the lens of academics working with the intelligence community. In doing so, I showcase a few underappreciated consequences of engaged scholarship. Lastly, I offer insights for policymakers to consider as they apply findings and reckon with the fact that they may not know everything about the provenance of the scholarship they consume. How research is produced, and with what motivations, should factor into their own ethical calculations about how best to use scholarship.

Engagement Within a Complex Policymaking System

> Policymaking is a complex system. It may seem linear and straightforward, but it hardly ever is. The process is a vibrant, cacophonous ecosystem of input, interactions, ideation, and iteration. As such, academic research findings and policy-engaged scholarship exist in this dynamic system, but they are but one of many contributions to policymaking—let alone outcomes of the policies. (Devermont and Steadman, 2021)

Policymakers know that well-researched evidence and rigorous findings help make them into better decision-makers (e.g., Avey et al. 2022). During my time in the US government, I witnessed many policymakers seek out research to improve their decisions. And this was not just to seek out specific findings to confirm previously decided matters. Sometimes they were driven by genuine curiosity, such as when I was asked to contact a scholar who studied Salafism in West Africa to better understand the connections between communities in the Sahel region and prominent students who had studied religion in Saudi Arabia. At other times, scholars shed new light on confounding circumstances. For example, a study of two Tunisian towns, similar in demographics and socioeconomics, explained why hundreds of young men from one town left to join ISIS, while from the other almost none did (Marcusa 2019). Scholarly evaluation of projects was also insightful, such as when research showed that government actors have biases that facilitate the continued funding of projects in the field that never deliver on their stated goals, yet somehow kept receiving funding. Policymakers sought out academics who were informed subject-matter area experts and looked to them for additional assessments of the likely effects, including downsides, of specific policies.

But academic expertise was rarely the only input, and researchers should be aware of this. For academics seeking to provide inputs to policymaking, knowing that the national security system values one's work is an important source of motivation. Researchers' appreciation of the complexity inside policymaking—that it produces various types of interactions and yields multitudes of impacts—can help them shape their involvement in policymaking, including taking precautions to ensure their own ethical redlines remain intact (see Lyall, Chapter 2, this volume). For instance, if a researcher knows that a decision has already been taken and proponents of that decision may be looking for justifying evidence to support it, a researcher might consider interacting with that policymaker in a different manner (e.g., perhaps withholding their input as a "rubber stamp" unless they favor the ethics of the policy outcomes; see Olsen, Chapter 3, in this volume). Conversely, if a researcher knows that a particular decision is fraught and subject to multiple internal policy viewpoints, they may choose to not only highlight their own findings, but also share contradictory results generated by other scholars (see Pielke 2007).

The complexity of policymaking means that research is not just inserted into a single point in the process or by one sole modality. For example, an insider to the policymaking process who arranges academic engagements can greatly shape which academics are invited or how insights are received based on questions asked or topics to be considered. With this type of engagement, for example, a scholar may or may not know whether their work is being presented as a panacea, as representative of an entire field of study, or as an outlying or disruptive finding. Alternatively, when research is brought into a policy process from the top (i.e., an email forwarded from a very senior official such as a General Officer in the military), it might be read with great interest and receive greater scrutiny than if shared by a less senior official. A scholar may or may not realize that they are engaging with a highly mobilized and informed audience, or alternatively one that is merely checking a box of having "engaged with academics." Some scholars may encounter ethical discomfort with engagements depending on whether they can anticipate how their work may ultimately be used, if the decisions may support (or undermine) policies they believe to be right or wrong, whether they will have an opportunity to shape policy, or if their input could potentially be a rubber-stamp or misused or dismissed.

Another layer of complexity stems from the policy influence of scholars' interlocutors possibly being invisible. Not all interlocutors, even officials who share the same rank or title, have the same influence over decisions. Influence might be based on internal structural considerations such as whether they control resources and/or personnel, or have the lead on a particular policy domain. It is not all about position, either, as policymakers with differing styles, internal reach, or informal power also shape their influence. A staffer who whispers closely in a senior official's ear is one archetype, while another is a staffer who rarely shares their scholarly engagements with other officials. A scholar's decision to engage may change based on who their interlocutor is and their placement, since this can affect their prospects for policy influence.

For instance, I witnessed scholars find themselves in live decisions at moments of high urgency. During the Somalia famine in 2011, scholars who had studied al Shabaab's taxation practices and economic export routes were able to provide a fuller picture of how US sanctions were impacting food delivery. One of the scholars had previously presented to a key staffer, who then invited them to share their insights at a dynamic moment in the process. Similarly, in 2010, a researcher presented findings on Yemen's long history of Houthi-led rebellions for several different Assistant Secretaries of State, yet each of them took away different implications for US–Yemen policy based on their institutional subject matters (i.e., counterterrorism, regional political affairs, democracy and human rights, political-military affairs; see Sisk, Chapter 4, this volume on how policymaker priors shape how they absorb research). Yet I also observed situations with foregone conclusions, and I wondered whether, had scholars been aware that key decisions had already been made, they would have changed the tone of their presentations or expected different outcomes from the interactions.

The more that researchers appreciate the nuances of how either nascent or fully formed a certain decision is, the better positioned they will be to decide whether and how to engage. Additional information researchers could seek out includes how a particular policymaker is or is not central to a decision, how challenges or other considerations might contribute to a decision, and the underlying politics of the issue at stake. Knowledge of these complexities can help the scholar decide whether and how to engage most responsibly so that they can be assured that their effort is achieving their expected outcome (see Olsen, Chapter 3, this volume). In short, researchers should become as informed as possible about the policymaking context before engaging, keeping in mind the complexities outlined above.

As research enters a dynamic policy process, policymakers with deep knowledge, experience, and expertise may have already reached conclusions about topics. Few policy deliberations are a *tabula rasa*, and no policymaker is an automaton that absorbs information with no prior conceptualizations (see also Sisk, Chapter 4, this volume). Furthermore, research findings may also be used differently by different bureaucratic players, depending on whether they are consistent with or disruptive of existing agency positions. Are agency positions already supportive of a piece of research, or will agency positions be called into question because of it? While some policymakers may be seeking to retain the status quo, others will try to upend it (see Olsen, Chapter 3, this volume).

This is all to say that academic engagement and specific findings enter a policy system with existing biases, positions, and bureaucratic incentives (see also Epstein, Chapter 7, this volume). Moreover, research and knowledge may have influence for reasons beyond the content of the research and beyond a researcher's intention. Awareness of these preexisting policy and power dynamics can help academics calibrate their perceived influence and possible ethical burdens of whether they may bear any responsibility for real-world outcomes (including in cases when their influence may be negligible). The more a scholar knows about the process and the point at which they are entering it, the better informed they can be about whether they are pushing on an open door, are being used as evidence ex post facto for a predetermined decision, or are contributing to live policy deliberations. Any of these scenarios may cause them to either lean in or pull back from engagement. With more knowledge about the process, scholars can decide for themselves how to engage. They can then make more tailored decisions to sit something out, opt for a public-facing or private engagement, or suggest others to join who might have something different to say or provide more evidentiary consensus.

At the same time, not all situations are complex. There are also engagements between researchers and policymakers that are relatively straightforward. Researchers are invited to brief policymakers on their work and asked for their assessments. They fill advisory roles in which they are asked to look at hard questions, and they contribute knowledge at conferences, events, or meetings where policymakers are present. Such direct engagement provides insight about the information entering the policy process. Scholars may get to know a specific policymaker, and perhaps even gain information about a specific decision. With more closeness and familiarity, researchers may have opportunities to ask questions about how their work is likely to be used. By receiving feedback about the process and at what point they are engaging, scholars can anticipate their likely impact as well as any ethical dilemmas that might arise.

Straightforward engagements can also yield candidness, interdependence, and trust-building. I have seen policymakers thank academics for their timely interventions and inform researchers about exactly how a piece of information was useful. I have helped draft high-level speeches that cite research studies. I have also watched as researchers were invited to speak at the United Nations, testified to Congress, or were called upon to comment as part of an advisory board. In these cases, scholars not only fulfilled a timely need on an important topic, but also fulfilled their own ethical imperative to share knowledge and participate in the court of public opinion. I have seen the government fund scholars to specifically study critical topics for future policies. Despite the tendency toward opacity in policymaking, not all decisions are made within a "black box." Opportunities for transparency, sincerity, and cooperation can enrich policy processes and improve policymakers' familiarity with research.

Nevertheless, even with strong incentives for policymakers to consult with researchers, there are weaker incentives to use research (or recommendations offered as part of the research) in the ways that researchers may intend. In some cases, the use of specific findings might be exactly as a researcher intended or recommended. I saw this happen almost overnight when studies began to show that job programs, by themselves, were not effective to deter youth from joining politically violent groups. But, in other instances, I witnessed that, despite broad readership of a study on the importance of youth empowerment and agency, it did not change the character of many of the government's youth programs since they were predetermined and not designed by the youth participants. In other cases, research may spur imaginative and novel ideas that neither policymaker nor researcher could devise independently.

Ultimately, understanding some of the potential ways that research is used inside policymaking can help scholars assess whether they are comfortable with certain types of policy engagement and help them ask the right questions of policymakers before deciding to engage. They can weigh what they know against what they do not and decide on the best path forward based on the situation, including the possible benefits or risks of not engaging. With more information, scholars can also inform policymakers about any ethical concerns they have and work with them to devise suitable mitigations to make the interaction more productive and to make policy more effective. A scholar who is concerned that their work will be overgeneralized, for example, might ask a policymaker if they can bring along some colleagues to provide a broader perspective. Or, a scholar who is worried that their work might be used to reinforce patterns of extraction and colonization can directly request that a policymaker include local partners (see Kaplan, Chapter 5, this volume). While not all situations will warrant this type of mitigation, open dialogue as a rule can engender mutually rewarding trust across the "gap."

The Mystery of Policy Engagement: Working with the Intelligence Community

At times, policymaking is not just complex, but also opaque. This opacity can have significant consequences for how academics assess their ethical and moral responsibilities: if they do not have visibility into how their work will ultimately be used (or not used), how can they decide how best to structure their engagement? While academics rarely have decisions solely attributed to them by policymakers, I have encountered many scholars who themselves feel responsible for their work being used to advance decisions that they did not support. In my experience, most of the scholars that I am describing here, who thought their work was used in ways that were disconcerting, only saw part of the picture. This is because many of the cogs of decision-making—including those that have nothing to do with a scholar's particular findings—are not public information. Nonetheless, when confronting opacity, it is easy to draw such correlations and make future ethical decisions to engage or not based on an incomplete picture.

In national security policymaking, engagement with the intelligence community (IC) is often viewed as opaque. In my experience, academic engagement with the IC is an automatic ethical redline for many researchers, which prevents them from engaging. While at the State Department and the US Institute of Peace, I met with several scholars who remarked very candidly that they were unwilling to engage with the IC because of their ethical concerns with the agencies' work. Such skepticism may certainly be warranted given the highly classified nature of much of the IC, the covert or clandestine nature of parts of it, its participation in many controversial historic decisions, and the reality that interlocutors may not be able to disclose their identities to researchers (Figure 9.1). I do not attempt to assuage those concerns, but rather seek to explain the role that intelligence analysis plays in policymaking.

Figure 9.1 President George W. Bush delivers remarks during the ribbon-cutting ceremony for the newly renovated White House Situation Room, May 18, 2007. The Situation Room is the hub for inter-agency crisis intelligence briefings and decision-making.

Courtesy Archival Operations Division, NARA

Intelligence analysis represents an indirect way that academic findings or scholarship may enter policy processes. In fact, decision-makers may never know that the information they consume has been shaped by academic engagement when it comes in the form of an intelligence briefing. IC-authored analysis may have exceedingly far reach as the IC's wide sharing of information spans countless agencies, and it can have deeper impact than engagement with any individual office or department. This cuts both ways when considering whether to engage, since it involves both extensive potential influence along with possible risks of harm, which may in turn depend on how a scholar decides under what conditions their findings might be most accurately used. But for complex decisions with multiple interested policymaking parties, IC engagement is a great way to ensure that the "knowledge picture" is more complete and includes academic findings that are appropriately contextualized.

As part of the national security process, the IC is charged with information collection, analysis, and creating shared understandings among different government agencies (The US Intelligence Community 2022). Intelligence often serves as the baseline for policy discussions, with finished intelligence products[1] providing the information landscape upon which decisions are considered. The most prominent, widely shared types of intelligence that are read by many officials across many departments and agencies are compilations—a combination of synthesis and analysis resulting from many distinct pieces of information. This is known as all-source[2] intelligence and is prepared by IC analysts who a) combine unclassified and classified intelligence, b) review intelligence sources for validity, c) make deductions and judgments, and d) provide a comprehensive snapshot of specific issues. Sometimes these analyses contain information pulled from publicly available sources, including published academic findings. Other intelligence analyses are entirely composed of classified information (The US Intelligence Community 2022).

Because of the nature of these IC products, policymakers may already benefit from academic information that forms the basis for intelligence judgments, even if decision-makers themselves never read or engage with academic source material directly. If an analyst reads, engages, or seeks

[1] Fused or "Finished Intelligence" refers to the combination and merging of information contained in various types of intelligence that has been filtered, probed, supplemented, and evaluated by intelligence analysts with deep subject-matter expertise on the topic.

[2] All-source intelligence often refers to information, gained from various intelligence disciplines, to include open-source (publicly available information), signals intelligence (derived from the interception of communications), human intelligence (derived from human sources), geospatial intelligence (derived from imagery and geographies), and other types of collected intelligence.

out academic material, the findings may make their way to policymakers circuitously. IC analysts are a key conduit through which academics and academic literatures reach policymakers as they rely on intelligence analysts as their most informed sources.

Additionally, the policy process digests information that is compiled and prepared by experts in the intelligence community. Implicit in this exchange is trust in and reliance on the analysis process. When a policymaker reads intelligence analysis, they assume the information was sourced properly, that methods of collection and analysis were sound, and that it was obtained and considered with the confidence and precision of a subject-matter expert. This, in turn, has bearing on how policymakers engage with scholars. I have seen policymakers replicate this dynamic in terms of showing high confidence in researcher processes, methods, and analysis. I saw firsthand how some policymakers view and process research in the same way as intelligence briefers, assuming fidelity and trustworthiness—that they are the experts. This inherent trust in expertise is beneficial when it accelerates interactions, facilitates discussion of important findings, and even improves policy outcomes. Such interactions also help scholars feel valued and gives renewed meaning to their work.

Intelligence analysis may also lead to possible misjudgments, misinterpretations, and harmful outcomes, as history shows. Intelligence products often come with confidence judgments and with structured analysis that allows it to assimilate many types of findings. If policymakers inaccurately assume that researchers have analogous methods, processes, or analysis, they may in turn infer that a scholar is presenting incontrovertible findings. Policymakers may not appreciate the importance of researchers exploring alternative explanations, possible upending scenarios, or external validity—that is, whether results apply beyond the specified context in which they were originally uncovered. For example, policymakers might assume a particular finding is more replicable, extrapolatable, or comparable than a researcher intended, especially if the policymaker in question usually engages with intelligence that makes such judgments. Thus, it can be helpful for researchers to understand this dynamic and, where necessary, point out how scholarship is different. Related to the problem of overgeneralization highlighted earlier in the volume (see Epstein, Chapter 7, this volume), researchers must be clear about study limits, possible upending scenarios, and alternative explanations. Policymakers would also benefit from learning more about how research is scoped, pursued, and carried out. They can then avoid making assumptions about similarities between research products and intelligence conclusions.

In sum, engagement with the IC can have several underappreciated ways of getting research into the hands of policymakers, even if at first glance it seems quite fraught. Some researchers avoid engaging with the IC directly because they believe that their findings may be used for spy-craft or poor decisions by the US government (something that I have heard from several scholars). Yet, by not engaging directly, scholars should also consider that their work will not reach many potential audiences across the government outside the IC who could benefit from its use. Moreover, since scholarly inputs are embedded within intelligence analysis anyway, scholars' work is still being used. This indirect means of shaping policy also has risks, especially because researchers have less ability to influence how their findings are used or interpreted due to the aggregated nature of IC products. Overall, it might not make sense for academics to avoid direct engagement with the IC since their work is influential in any case. Rather, direct engagement might provide opportunities to improve the uses to which research is applied.

Making Policymakers More Responsible Consumers

When discussing "responsible engagement" and how to improve the transmission of research findings to policy implementation, researchers are only one half of the equation. Much work remains to inspire policymakers to take their roles as consumers of engaged scholarship more seriously and responsibly (see also Olsen and Sisk, Chapters 3 and 4, this volume). While policymakers know that they are accountable for their ultimate analysis and decisions, they may not have sufficiently reflected on how they might more responsibly use research to inform their decisions. In my experience, policymakers are not fully informed about the research enterprise or the ethical dilemmas and full range of incentives that academics face as they approach policy engagement.

As one example, when I was at the State Department I benefited enormously from a historian who had studied al Shabaab and provided important insights about the group and the longer history of political Islam in Somalia. However, I never asked how their research was funded. In retrospect, I might have thought differently about the research depending on whether it was funded by the US Department of Defense, the government of Qatar, or an anonymous financial donor. Also, with the benefit of hindsight, I might have raised different and more probing questions about the scholar's past work, their incentives around publishing, and the inclusion of coauthors of differing stature. I might have asked more about their confidence in the findings, how much was field-based versus desk research, and about some of the protocols concerning how conclusions were reached. Perhaps none of these questions would have changed the analysis or my interpretation of it. But I cannot say for certain, because at the time I did not know enough about academic research and the professional incentives that may have influenced the presentation of findings. Individuals in policy positions with no formal research training may not appreciate the ways in which funding, methods, and theories can impact the scholarly conclusions on which they rely.

On the policy side, officials should take responsibility for understanding ethical dilemmas that may be hidden inside academic engagement or policy-relevant research. It is not enough just to use, fund, or rely on academic findings. Policymakers could be more proactive and ask more questions of academics. They could improve their own literacy regarding the academic products they are using to understand the limits of their possible uses, to appreciate the incentives that might have shaped the work, or to otherwise consider the inherent aims of the scholar that may be either consistent or inconsistent with the policymaker's aims. Whether because of a lack of familiarity, lack of time, or lack of interest, there is not enough attention from the policy side to the "sausage-making" of research. This includes everything from timelines under which studies are conducted, the applicability of findings to other cases, or emerging methods that are relatively untested and not yet mainstream. All may have relevance to the ultimate findings and the ways in which policymakers can or should consider scholarship.

With more research literacy and knowledge about engaged scholarship, policymakers could start to recognize the risks of misuse of research, scope out bias and prior political commitments, recognize divergent interests, and become more comfortable with partial answers and complexity. Earlier I referred to the study of two Tunisian towns, a project that was widely read and even used as an example in the US Institute of Peace's "Task Force on Extremism in Fragile States" (Task Force on Extremism in Fragile States 2019). The study was explicit about the findings being localized to the two Tunisian towns, and at the time it had not been replicated elsewhere. However, some policymakers were quick to see the findings as generalizable to other towns, countries, and even globally. But those policymakers would have been wise to heed the researchers' cautions, and to consider some of the limitations of the study and the localized context before assuming broader applicability.

To be sure, academia can be opaque in its own ways. Policymakers would serve themselves well by asking more questions that uncover the opacity or complexity of the academy (or, similarly, research pursued by interest groups, the private sector, independently funded researchers, or the like). To understand the broader context in which research findings are making their way into policy processes, policymakers could ask critical questions, including the following:

- Who funded the research, and did the funder influence the specifics on what to investigate?
- What prior work did a scholar rely on, and is that a representative sample of the field?
- Are the findings disruptive to the field and are there known incentives/disincentives to maintain stasis or to disrupt?
- Is the work replicable? Are the findings broad and repeatable or are they deeply specific to the context and time in which they were studied?
- What safeguards were put in place for research subjects and investigators when conducting the work (i.e., was anyone put at risk in the conduct of the study)?
- Was the research participatory and did the investigators work with communities to share findings?
- How did the research consider recency or historic biases?
- How much has changed in the intervening time since the research was completed, and should one still have confidence that the findings are reliable?
- Why were some methods chosen over others, and are peer-reviewed publications a result of the research?

These questions (and many more) would represent a commitment to a level of research literacy that would make policymakers more conversant in the opportunities and limitations of research. This would not only help to avoid misinterpretation, but also encourage conversations that allow policymakers to assess the applicability of a study to a policy decision. It also could chart a path in which decision-makers and researchers are, together, able to more fully interrogate the research they are respectively consuming and producing, as well as its ethical implications. Engagement that includes a scholar who can ask a policymaker about how their work is likely to be used, combined with a policymaker who can ask a scholar about replicability or recency bias, is sure to lead to more responsibly informed policymaking.

Conclusion

Overall, the case studies in this volume provide close accounts of engagement dilemmas that researchers often encounter. These include trying to adjudicate competing findings for policymakers, reporting unpopular findings, publicizing one's ethical redlines, avoiding overgeneralization or overstatement, and grappling with one's own positionality vis-à-vis a research subject. My own reflection reinforces the importance of those insights, but also points to an additional consideration, which is for researchers to know as much as they can about the specific point in the policy process when their engagement takes place. Policy processes are often highly complex and dynamic, and many factors figure into a decision other than research findings. However, this is not to dismiss the importance of scholarly work or its potential to productively impact decision-making. Rather, my main argument is that academics' engagement dilemmas can be mitigated by careful scrutiny of the engagement context ahead of time. The chapters by Olsen (3) and Sisk (4) on presenting unpopular findings and interpreting competing research claims, respectively, provide particularly salient cases illustrating the importance of understanding the engagement context.

Through deeper knowledge of policy processes, researchers can avoid overstating but also underestimating their policy impact. Notably, I have encountered many researchers in my career who take principled positions against direct engagement with the IC. That is certainly a legitimate decision. Yet those researchers should also recognize that if their work is in the public domain, it is very likely making its way into intelligence analysis regardless of their commitment to avoid direct engagement. The question then becomes whether it is indeed better to avoid direct engagement or, rather, to avail oneself of the opportunity to contextualize findings and discuss their limits and applicability with the powerful actors who actually shape policy.

Finally, considering my own policy work and the multiple policy contexts in which research has played a part, this reflection also serves as a call to policymakers to become better informed about academia and the research enterprise. Greater research literacy among policymakers would help them be more aware of the importance of scholarship that explores alternative explanations and study limits. It would also make policymakers more sophisticated consumers of research if they knew who was funding the work, what incentives academics encounter to engage or to stay away from policy work, and whether there are political biases built into particular theories or methods. Deeper mutual understanding between the research and policy worlds can be achieved through sustained engagement and asking probing questions. This can only help improve the transmission of academic expertise to policy processes.

References

Avey, Paul C., Michael C. Desch, Eric Parajon, Susan Peterson, Ryan Powers, and Michael J. Tierney (2022) "Does Social Science Inform Foreign Policy? Evidence from a Survey of US National Security, Trade, and Development Officials." *International Studies Quarterly* 66(1): 1–19.

Devermont, Judd and Leanne Erdberg Steadman (2021) "The Relative Importance and Unimportance of Academic Contributions to the Policy Process." *Duck of Minerva*, July 6. https://www.duckofminerva.com/2021/07/a-piece-of-the-puzzle-the-relative-importance-and-unimportance-of-academic-contributions-to-the-policy-process.html.

Marcusa, Michael (2019) "Radicalism on the Periphery: History, Collective Memory, and the Cultural Resonance of Jihadist Ideology in Tunisia." *Comparative Politics* 51(2): 177–197.

Pielke, Roger A. (2007) *The Honest Broker: Making Sense of Science in Policy and Politics*. Cambridge/New York: Cambridge University Press.

Task Force on Extremism in Fragile States (2019) "Preventing Extremism in Fragile States: A New Approach." *The United States Institute of Peace*. February 2019. https://www.usip.org/sites/default/files/2019-02/preventing-extremism-in-fragile-states-a-new-approach.pdf.

The United States Intelligence Community (2022) "How the Intelligence Community Works." https://www.intelligence.gov/how-the-ic-works.

The Commission on the Intelligence Capabilities of the United States Regarding Weapons of Mass Destruction Report to the President of the United States (2005) "An Intelligence Primer: Appendix C." https://www.govinfo.gov/content/pkg/GPO-WMD/pdf/GPO-WMD.pdf.

10
The Future of Responsible Engagement

Rachel A. Epstein and Oliver Kaplan

A common view of academics is that they are trapped in the ivory towers of their university campuses, physically and intellectually separated from the "real world." This separation from pressing day-to-day events can be beneficial for producing deeper analyses and insights. But, increasingly, scholars are encouraged—exhorted—to "be useful" and make sure their research is "relevant" to the issues of the day (Nincic and Lepgold 2000; Walt 2005; Desch 2019). They may also have careerist incentives to be relevant, whether to gain exposure for their work, because of new "policy impact" criteria for tenure and promotion or departmental rankings, or to be seen as public intellectuals. In the pursuit of relevance, scholars may have idealized notions of research and engagement, where research is clear and policymakers run with their findings and advice. It conjures a simple formula: bridge the gap, improve humanity, maybe even reap some professional glory.

But policy interactions rarely play out like that. We do not always know ahead of time what studies will show or how findings will be used or viewed by whom. That's science. A "gap" remains between academics and policymakers, and "bridging" it does not automatically translate into successful, informed policymaking. Academics, from their perch, have a point of view: about reaching for the "right answers" or the "truth" and about how findings should be used. For their part, policy actors also have their own views, priorities, and incentives, based on where they sit, their issue portfolios, as well as their values and even politics.

Should scholars impart the full "truth" in their findings in all circumstances? Or should they only share insights that will be welcomed by practitioner audiences or that they feel will not risk harm? There are many considerations. This volume does not seek to deter scholars from bridging with policymakers and the public—far from it. Rather, it seeks to raise awareness about the options and risks for policy-engaged researchers as they pursue

Rachel A. Epstein and Oliver Kaplan, *The Future of Responsible Engagement*. In: *Speaking Science to Power*.
Edited by: Rachel A. Epstein and Oliver Kaplan, Oxford University Press.
DOI: 10.1093/oso/9780198875185.003.0010

their engagements. In this sense, what we are doing here is different from standard "bridging-the-gap" exercises and the quest for relevance. "Bridging" skills include practical tasks, such as how to organize presentations (e.g., bottom-line-up-front and elevator pitches), how to communicate (use accessible language/avoid jargon), and how to write op-eds and give briefings. The skills also include how to assess policymaker motivations and recognize that they may have short availability and attention spans and lack deep expertise on some issues.

An imperative of the push for "bridging" might be characterized as to always do more bridging. And yet, in a recent survey almost all International Relations scholars who responded agreed that should their recommendations be adopted by policymakers, they "bear at least partial responsibility for the impact of those policies in the real world" (Hendrix et al. 2023). So, we offer a related imperative: be responsible. Going beyond the "how" of "bridging" the gap with policymakers, we have sought to ask thoughtful questions and suggest tools to encourage *responsibility*—for researchers both to protect themselves and to optimize the contributions of the knowledge and findings that they are sharing. Drawing on the deep reflections by our practice-oriented group of social scientists, we have documented some of the main policy engagement dilemmas that can arise.

Some elements of responsible engagement that we articulate have been described by previous authors, organizations, and fields, albeit perhaps by different names. There are references to engagement dilemmas in pop-culture, by organizations with long historical legacies of cautioning on science-to-policy, and in fields such as science communication and philosophy of science. In one of the most accessible examples that we discussed in the Introduction, the physicists Robert Oppenheimer and Joseph Rotblat were tormented by ethical quandaries in the quest for the atomic bomb. There are also current parallels surrounding many research-to-application interventions affecting daily life, including the invention of the internal combustion engine and the unforeseen downstream impacts of its carbon emissions on climate change, the benefits of opioid pain-killers and the risks of dangerous and addictive narcotics such as OxyContin or Fentanyl, or the raucous debate over the promise and perils of artificial intelligence (AI).[1] Many scholars have no doubt been aware of some of these dynamics, even if they had not put a name to it. We did. By introducing "responsible

[1] In an eerily similar lament to that of Oppenheimer, the so-called "Godfather of AI," computer scientist Geoffrey Hinton, expressed sentiments of regret and concern about the potential harm from the technology he helped to develop, saying, "I console myself with the normal excuse: If I hadn't done it, somebody else would have" (Metz 2023).

engagement," we note that while there are many boons from bridging the gap, the potential perils also loom large.

We therefore also proposed some possible approaches to responsible engagement. We have highlighted a set of questions and principles to consider for academics and policymakers to responsibly adapt to new situations—and at each step of the research-to-policy process (Table 10.1). In addition, the volume suggests general strategies and some specific solutions for avoiding the pitfalls of overgeneralization, the misuse of research, and problems of positionality and power imbalances. These guidelines can inform policy activities in International Relations and political science, sociology, anthropology, economics, and history, but also in fields such as environmental policy, climate science, technology, biomedical science, and public health, among others. "Responsible engagement" can also inform the actions of nonacademic researchers who hold positions in think tanks, NGOs, or government agencies.

Being responsible in policy interactions can entail bearing costs, at least in the form of time, thinking, planning, and reflecting. It can also mean foregone opportunities for research support or influence, especially if responsibility means losing valuable relationships with policy actors. It could even mean walking away from projects completely and accepting the sunk costs that entails. Yet César Caballero could bear the personal physical safety risks of walking away from his government post, as others in similar positions have done. If he could do that, it should not be too much to ask of scholars to take the extra time and precautions needed to avoid possible policy engagement harms.

In this concluding chapter, we review the common "engagement dilemmas" we identified and explore their implications for scholars as well as donors and policymakers. We also offer a critical view of our cases, examining

Table 10.1 Responsible engagement general principles

- Recognize that researchers have control over some aspects of engagement, but not all
- Practice introspection regarding the foundations and scope of one's claims
- Anticipate potential harms when providing policy advice and work to minimize harm
- Build relationships and communicate with policymakers
- Articulate ethical redlines prior to conducting research
- Follow formal disciplinary norms and challenge potentially damaging informal practices
- Be inclusive of different policy stakeholders
- Plan for worst-case scenarios
- Learn from experience, learn from others, and seek advice
- Be transparent about political commitments and potential conflicts of interest

what could have been done better with the aim of learning from our initial lapses. We call on actors on both sides of the “gap” to engage in better future practice to achieve more productive and equitable policy outcomes. We also call for a wider research agenda concerning the study of responsible engagement.

Policy Engagement Dilemmas and Researcher–Policymaker Relationships

As a group of scholars struggling with how to share our research, our core contribution has been to identify an initial set of “engagement dilemmas.” We covered the issues of unpopular findings, prosocial lying, overgeneralization, e-hacking and ethical redlines, positionality and inclusion, and competing claims. Although the dilemmas were studied in relation to specific topics and in specific contexts, we believe they are likely to be encountered frequently by scholars in many fields, working on many topics, and interfacing with diverse types of policy interlocutors, from national leaders to NGOs and local communities (Sanford and Angel-Ajani 2006).

The dilemmas arise because of the nature of the interaction between researchers and policymakers—the consumers of their research products—which can resemble “principal–agent” relationships (e.g., Ross 1973; Mitnick 1975; Jensen and Meckling 1976). In these relationships, a principal contracts an agent to complete a task, but they may have diverging incentives and the principal cannot perfectly observe the agent’s performance, which can lead to shirking. In a first type of principal–agent relationship related to “bridging the gap,” scholars can sometimes be thought of as the “principals” who in a sense provide instructions to the policymaker “agents” through their findings (though they do not contract policymaker agents for a task or exert direct control).[2] But by sharing their insights, scholars usually seek to achieve certain goals and make things better. With the possible diverging incentives between the two “sides,” however, the researcher “principals” may prefer to put some boundaries or guidelines on how the evidence might be used by the policymaker “agents” and contribute to outcomes (or avoid outcomes deemed unacceptable, such as certain harms or inequities).

Two examples from the case chapters speak to this first type of interaction. As a researcher, Lyall (Chapter 2) was sought out by military officials

[2] It is not precisely a classical “principal–agent” relationship, since researcher “principals” are not dictating directives to policymakers that they expect to be carried out by the policymaker “agents.” Nonetheless, researchers do seek to provide information that helps to better guide the policymaker’s hand.

for advice. Yet he had concerns (as a "principal" sharer of research) that the military officials who were the recipients of his research (i.e., "agents") might use his analyses of the effects of airstrikes in Afghanistan in unintended and possibly morally suspect ways (and he might have been challenged to observe such uses).

Similarly, as a government official responsible for generating evidence and data in Colombia, Caballero Reinoso (Chapter 8) hoped to act as a "principal" and inform good policy. But he also faced worries, like Lyall, when he presented results to President Uribe and his staff, as they did not like some of his conclusions and put pressure on him to change them.

It is also possible for a second type of relationship to exist: one in which the policymaker becomes a "principal." In this case, the policymaker attempts to curate the research inputs they desire from scholarly "agents" for a policy decision. They seek out particular academic perspectives—to access expertise or affirm prior decisions or preferences. It is conceivable that Olsen was initially contacted by the UN Working Group (UNWG) on Business and Human Rights with the aim of having her furnish particular findings and viewpoints that were of interest to the working group staff. As Erdberg Steadman also explains (Chapter 9), in some circumstances policymakers may have very clear ideas about the kinds of evidence they are seeking.

As in standard principal–agent relationships, policymakers and researchers can have differing incentives, motivations, and positions. Researchers want to be heard, have influence, and make sure the right information reaches the right policymaker and is used effectively. But researchers may also elevate their own findings and agendas, eliding others (when acting as Pielke's [2007] undesirable "stealth advocates," for example). Policymakers' interests can also vary. They may seek to make objectively optimal decisions, but their perspectives on what matters and what is urgent can vary (Neustadt 1960). Where they stand on a given issue is partly based upon where they sit: to whom they report, what issues take priority over others, or the politics of the day. In some situations, researchers and policymakers may have shared interests in addressing a policy issue, and the risks of the misuse of evidence may be low. In other cases, researcher and policymaker understandings and interests may diverge.

The two actors in the relationship have different information about the policy process, the policy problems, and politics. From the perspective of researchers, the policymaking process can resemble an opaque "black box" into which they may have little visibility. They may enter prior to a decision, in the middle of a discussion, or after a decision has been finalized. They may also be called upon to provide input to one side of the debate or another. Either way, academics are challenged to observe how the policymaker intends to use their research. For instance, Kaplan and Serna (Chapter 5) had little initial notion about how their briefings on community-based atrocities prevention would be used (and only gained better insight long after the engagement, when the survey results of the policy participants were shared back with them).

For their part, policymaker "principals" seeking insight and advice may not know the scope of the available academic research. With perhaps relatively less expertise on an issue, they may not know whether they are being presented with the scholarly consensus on an issue (or competing claims) or are instead interacting with "stealth advocates" who share only partial perspectives (Pielke 2007). It is this basic misalignment of incentives and information that gives rise to many of the engagement dilemmas that we have elaborated, as well as to some of the remedies. In sum, the principal–agent framework is helpful for assessing engagement relationships, but these relationships can be complex and even bidirectional, with at times unanticipated shifts in roles.

These engagement relationships can be vividly analogized according to Vincent Van Gogh's famous painting, *The Starry Night*.[3] The painting is an abstract composite of cascading celestial swirls of bright stars and the moon amid a dark blue night sky. Different scenes play out in different parts of the painting, and how a viewer interprets the meaning of the painting may vary greatly depending upon to which part their attention is drawn. From afar, taking it all in, one sees the gestalt that Van Gogh perhaps intended—a generally dark night sky with a few points of light—perhaps representing a complete if gloomy image or assessment of policy. However, if zooming in on the moon, the night sky falls away, and all that comes into view is a bright yellow object, suggesting perhaps a rosier policy conclusion (see Figure 10.1).

The principal–agent-like dynamic plays out across the policy space of the painted canvas. Some principals (researchers) may steer agents (policymakers) toward one part of the painting or another in the way they present their findings. Or some policymaker agents, seeking particular outcomes—and information or perspectives to support those outcomes—become "principals." They curate particular researchers (or particular information) to focus on the areas of the painting they care most about. For example, in the diagram in Figure 10.1, a scholar could steer a policymaker to their own research, perhaps shifting their view toward the moon. Or, if policymakers want to ensure that an optimistic view is seen, they may look for researchers who can tell them and wider audiences to focus on the bright spot of the moon. If the

[3] We thank our colleague and friend Cullen Hendrix for this insightful formulation.

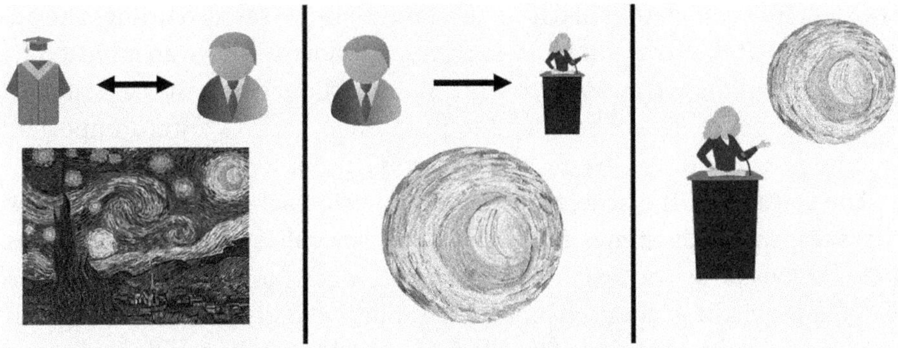

Figure 10.1 Vincent Van Gogh's *The Starry Night*, with some "responsible engagement" modifications (courtesy of Cullen Hendrix). The two-way arrow between scholars and their policymaker interlocutors reflects that each can become either "principals" or "agents"—to provide particular advice or curate particular advice—depending on the circumstances.

aim is for policy judgments to be based on a comprehensive view of available evidence, however, the principal–agent-like interactions may obstruct sound decision-making.

Rory Stewart, a former British Secretary of State for International Development and Member of Parliament, astutely depicted an example of this *Starry Night* dynamic. In his various roles as both an advisor/expert (though not academic) and a policymaker, he reflected on the debates surrounding policy toward Afghanistan and Pakistan, on which he advised given his expertise. He recalled how experts may be selectively exploited by policymakers. He provides the analogy of wayward policymakers' possibly ignoring the urging of experts to focus on a top-level decision—such as whether to drive a car off a cliff or not—and instead focus on a subordinate question, such as whether to wear a seatbelt. As Stewart concluded:

> It's like they're [policymakers] coming in and saying to you, "I'm going to drive my car off a cliff. Should I or should I not wear a seatbelt?" And you say, "I don't think you should drive your car off the cliff." And they say, "No, no, that bit's already been decided—the question is whether to wear a seatbelt." And you say, "Well, you might as well wear a seatbelt." And then they say, "We've consulted with policy expert Rory Stewart and he says...." (Stokes 2009)

In Stewart's anecdote, an expert's findings could be selectively adopted to support a particular outcome. Academics can become part of a mere "box-checking" exercise to affirm already-decided policies that they may disagree with, and into which they had little substantive input (see also Erdberg Steadman, Chapter 9, this volume). In such cases, academics have an ethical duty to be aware of these dynamics, to work to understand where they are situated within the policy "black box," and to push policymakers toward important questions and findings, even when inconvenient for them.

The severity of the scholar–policymaker principal–agent dynamics and the engagement dilemmas may vary along several important dimensions. One dimension is the stage of one's career. More junior scholars may not feel they have the academic record or confidence to break away from policy engagements if they go sour. The dilemmas may also play out differently from country to country and context to context. For instance, in some settings, scholars may have much more authority or insight relative to policymakers, whereas in others, policymakers may have many options for scholarly input. The ability to have research-based influence may also depend on the policy domain and one's field of study and approach, as policymakers have voiced different perceptions of the informative value of different fields, with national security officials, for example, especially prioritizing history and areas studies (Avey and Desch 2014). There are also multiple forms of policy engagement that can occur at multiple stages of a policy or decision-making process. Whether co-creating research, participating in informal conversations, contracting as a policy consultant, or briefing findings after-the-fact, some interactions can involve more risks and more duties than others. The key point is that in any of these cases both researchers and policymakers have *agency*—they have choices and options to behave responsibly.

Guidance for Scholars to Engage Responsibly

We develop suggestions for responsible engagers at each step of the research-to-policy process. Most of the guidance comes from the engagement dilemmas, though some comes from our broader experience and additional examples. In developing our guidance for managing these dilemmas, we contemplate what the academics and/or policymakers could have done better in the case chapters. Some involve the general advice of simply gaining awareness of potential dilemmas and learning how to anticipate and identify them. Scholars can learn to think ahead about possible strategic situations from the research phase through to results-sharing. They can envision best- and worst-case scenarios. Other steps take the form of specific recommendations for hands-on actions, either more introspectively or in direct interface with peers or policymakers. Scholars can engage in self-assessment and establish and share their ethical redlines. They can reflect on how to educate policymakers so they are more aware of researchers' capabilities, incentives, and concerns.

Table 10.2 lists key questions that researchers may want to ask across the phases of their research and engagement. Many of the questions arise at the pre-research or research execution phases—perhaps *far earlier* than some scholars might be used to thinking about policy engagement or their responsibilities. Indeed, if a scholar is asking some of these questions "in the moment" of ethical doubt, they are probably already behind the curve. Other questions are more relevant and important to ask during the post-research phase, when the sharing of findings most often occurs. Our key pieces of guidance for responsible engagement are listed in Table 10.3.

X-Ray the Policy Black Box

A good starting place for approaching policy relationships is to X-ray the policy "black box." Scholars should work to gain greater visibility into how the policy process they are facing works and the possible risks and rewards. For some scholars, the analogy from game-theory of backward induction may ring true: to look ahead to what the policymakers may do given their new-found scholar-provided knowledge, and then prepare from there.

Table 10.2 Key responsible engagement questions

- Where am I entering the policy process? Am I part of a box-checking exercise?
- How much information should I share, and with whom?
- How will policymakers react if I find one result versus another (if the findings are unpopular)?
- What are my ethical redlines?
- To whom am I accountable if I cross my ethical redlines?
- Am I framing my project and findings for prosocial outcomes or possible malintent?
- What would happen if my research ends up in the hands of sinister actors?
- Am I clarifying competing claims and providing context-specific advice?
- What is my positionality vis-à-vis research participants and policymakers?
- Am I including research subjects or stakeholders? Am I representing their views?
- Am I being fully transparent concerning the limits of knowledge and findings for potential policy consequences?
- Am I being accurate about the applicability of findings, or am I overgeneralizing?
- How can I learn about the policy process?
- How can I get feedback about the effects of my inputs into the policy process?
- Which individuals, groups, or sectors might benefit more than others?

Table 10.3 Responsible engagement guidance

Phase	Dilemma	Guidance
Pre-research	*Unpopular findings*	Build positive relationships and communication with research sponsors
	E-hacking and redlines	Establish and share redlines; identify conflicts of interests, anticipate pressures, set up processes for accountability
		Assess policymaking context/"black box"; with whom to engage or not; what topics to work on
		Envision best- and worst-case scenarios
During research	*E-hacking and redlines*	Get advice from mentors and peers; make use of available resources that address ethical questions
	Overgeneralization	Explore alternative explanations, consider study limits and the implications of most likely upending variables
	Prosocial lying	Identify and confront harmful informal disciplinary norms (e.g., prosocial lying by academic or policy experts)
Post-research	*E-hacking and redlines*	Review whether any ethical redlines may have been crossed and redraw them as necessary
	Positionality	Reflect on one's own position; listen to stakeholders and include their views
		Hold-up sharing research, or consider a partial hold-up
		Avoid negative framing of research to avoid malintent
		Frame positively and provide disclaimers by specifying ethical uses and risks for specific stakeholders
		Get feedback from stakeholders, peers, and mentors
	Overgeneralization	Critically review scholarship for quality and accuracy of predictions and implications, especially as new topical events unfold
Always	*Prosocial lying*	Teaching and training on academic norms; educate policymakers
	Overgeneralization	Teaching and training on academic norms in the service of precision, transparency, and circumspection
	Competing claims	Tailor research, adjudicate competing claims, clarify trade-offs, contextualize findings

One can peek inside the policy black box by considering a few key questions. A first question to ask is: Who is the policy actor? Where are they situated within a bureaucracy or organization, what are their incentives, and what influence do they have over outcomes? This can provide insight into what a key interlocutor might do with findings and what degree of control they have over the policy process. Second, when in the policymaking process is a researcher intervening? Early briefings might provide more general, agenda-setting knowledge, whereas briefings that are conducted mid-decision or mid-crisis might be highly influential (as noted by Erdberg Steadman [Chapter 9]). Briefings post-decision might only be used for justificatory purposes. This leads to the final question of what policymakers' motivations are for seeking input. Is a scholar bringing desperately needed information that was previously unknown? Or could a scholar's input merely be part of a "box-checking" exercise—entering after a decision has been made simply to signal that a scholar was consulted?

The answers to these questions can determine what temptations might exist for researchers to deviate from their ethics or whether policymakers may be using scholarly input for ulterior motives. If scholars can obtain answers, they can responsibly adapt their approaches. One way to get answers is to get input and feedback from policymakers either before, during, or after policy interactions about what policymakers are seeking, what kinds of findings are valuable, and how findings are or were used in the decision-making process. Building positive scholar–policymaker relationships can help with obtaining policymaker feedback.

Building Relationships and Communication with Research Sponsors

A key risk of conducting policy-engaged research is potential conflicts of interests, in which policy actors may provide scholars with resources or public exposure, but with the expectation of receiving particular findings. As Tricia Olsen's experience shows (Chapter 3), without a meeting of the minds between scholars and policymakers, applied research can end up generating "unpopular findings" that are quickly discarded or ignored. Furthermore, policymakers often look to ground their decisions in a scholarly consensus—when it exists (Avey and Desch 2014). Sisk (Chapter 4) shows that it is incumbent on scholars to help clarify competing claims when possible so policymakers have clearer guidance from scholarship.

Responsible scholars should pursue clear and early communication with policy interlocutors—in the pre-research or pre-briefing phases—so that findings are not blankly discarded nor relationships tarnished (as also encouraged by Barma and Goldgeier 2022). Researchers should ask how policymakers might use particular findings if they arise from the research process. With better and earlier communication, Olsen might have decided not to get involved with the UN's business and human rights project at all.

Or, she might have reached a clearer understanding with the UN agency from the start about what findings might have been unpalatable or whether it was too late incorporate new information. She could have then decided ahead of time which aspects of the project to continue working on and which to set aside. According to Sisk, scholars adjudicating the findings on power sharing could have done more to signal in which contexts particular models and solutions were most likely to succeed for sustaining peace based on case histories.

Avoiding Seduction by Sponsors and Setting Redlines

Most policy engagements come about and begin with cordial relationships between scholars and researchers, and hopefully they continue that way. Yet even when engagements start on the right track it is possible that they may later get derailed by possible diverging interests and incentives or the emergence of new scenarios and conditions. As Lyall warns—and as Caballero Reinoso (Chapter 8) experienced—in the heat of an engagement there can be both temptations ("seduction") and pressures to possibly make ethical compromises that affect policy outcomes or scholarly integrity and reputations. Indeed, policy actors may dangle various glittering objects in front of scholars, such as opportunities for funding, data collection, or policy access and influence. Lyall and Caballero might have done more to plan ahead and set clearer redlines. That way, when their interlocutors made what these researchers felt were unreasonable demands, it would not have come as such a shock.

This risk of "e-hacking" (ethics-hacking) to preserve relationships with policymakers can be addressed by several steps of reflection. Scholars should first recognize that there can be sunk costs from getting invested in a research project and collaboration, and it can be hard to walk away, even if that is what one must do. For this reason, responsible scholars must also assess their situations and own motivations, to identify potential conflicts of interest and anticipate possible pressures from policymakers. They should then articulate and share their "redlines"—the ethical boundaries they are not comfortable crossing. This involves first imagining what kinds of potentially compromising situations or requests might arise. Then, ideally, it involves articulating and sharing them with peers as way of remaining honest and committed to them. Teams of researchers may also need to ponder and articulate both their individual and their collective redlines to come to a consensus before a project commences.

How to Share Findings: Affirmatively Framing and Caveating Research

Many scholars have had concerns that their research could be misinterpreted or, worse, misused for problematic purposes. This has been termed the "dual-use problem" (Akoto 2020). Once our research leaves our clutches, is published, and becomes public, there is no going back. It is not necessarily obvious who will later end up consuming research or where their interests may lie. Scholars may have visions of best-case scenarios when policy-engaging: of research making its way to important decision-makers who are looking to become informed to make the best prosocial decisions possible—we are often guilty of this ourselves! Some might aspire for their work to land on the bookshelves and book lists of President Barack Obama, Oprah Winfrey, Bill Gates, or some other influencer. And it is indeed wonderful when someone's work ends up there.

But there are some more problematic contingencies of which researchers must also be aware. What if, instead of ending up on a respected leader's bookshelf, scholarship ended up on the bookshelf of someone far more sinister, such as Osama Bin Laden? Unrealistic? Hardly. Upon raiding Bin Laden's compound in Pakistan in 2011, US intelligence agencies found numerous current-events and scholarly books (Office of the Director of National Intelligence 2015). In a similar story, the USB stick of the former leader of the United Self-Defense Forces of Colombia (AUC) paramilitary group was obtained by Colombian intelligence services, with contents indicating he had read various scholarly books, op-eds, and other published sources (Semana 2008). Shortly after the publication of *Why Nations Fail: The Origins of Power, Prosperity, and Poverty* by Acemoglu and Robinson (2012), a prominent news story in *The New York Times* about the civil conflict in South Sudan reported that the text had made its way to Riek Machar, former vice president-turned rebel leader (Fortin 2014).[4] Depending on one's point of view, an actor such as Machar might have been drawing from the research (or using it as justification) not to prevent a nation from failing but *precisely to make one fail.*

Given the possible risks of misuse of research, scholars should first do their best to make it *less* easy for their work to end up in the hands of malevolent

[4] "He has a satellite phone and a shiny touch-screen tablet in a battered brown case, and in his free time he is working through a paperback copy of 'Why Nations Fail.' The book, he says, inspires him to reflect on whether he is making the right decisions, though he holds Mr. Kiir responsible for South Sudan's bloody conflict ... Of course, rebel forces have been implicated in widespread civilian deaths during the course of the war as well, with the United Nations reporting 'mass atrocities' committed by both sides."

actors or to be exploited by them. One option for limiting nefarious use comes down to framing, which can even begin with the titles of scholarly works. There can be strong incentives to frame scholarship in flashy ways that draw interest from wider audiences. However, sometimes the framing of scholarly works can inadvertently invite misuse. Though tongue-in-cheek, some works have been framed in terms of ethically dubious outcomes, including such titles as *The Dictator's Handbook*, *When Violence Works*, "How to Win Friends and Wage Jihad," *A Field Guide to White Supremacy*, "Give War a Chance," *Coup d'état: A Practical Handbook*, *Making the World Safe for Dictatorship*, or *Rules for Rebels*.[5] To be sure, we *do not* mean to say these or other works have done harm and, indeed, dangerous actors could cause harm regardless of research framings. Yet, there is evidence that negative framings on political issues can have deleterious effects on assessments and attitudes (e.g., Carpenter et al. 2020). Scholars should be aware that the subtlety of their claims and works could be lost on some potentially bad actors, and even be viewed as handbooks for them. So, why make it easier for them?

Fortunately, there is an alternative approach of framing projects and findings affirmatively. That way, at least on the book cover or in the paper title, the focus is on the upstanding consumers of research who are encouraged to apply insights benevolently. Moving beyond the more visible titles of works, scholars can also provide specific guidance to specific types of policy actors within their texts. They can transparently acknowledge the benefits and risks of policy implementation. They can also include relevant policy- and application-focused caveats for findings, thereby reducing the chances of misunderstandings and misuse.[6]

Sharing Findings and the "Hold-up Problem"

Once scholars decide to "engage," they must confront the questions of how much to share and with whom. Scholarly publication and engagement can

[5] See Dukalskis 2021; Barron 2019; Luttwak 1968, 1999; Belew and Gutiérrez 2021; Gartenstein-Ross and Koduvayur 2019. According to *Rules for Rebels* (Abrahms 2018), "successful militant leaders have followed three rules." *The Dictator's Handbook* (Smith and Bueno de Mesquita 2011) description justifies the analysis it presents by observing that "The picture the authors paint is not pretty. But it just may be the truth, which is a good starting point for anyone seeking to improve human governance."

[6] The concluding chapter of Kaplan's book (2017: 310) seeks to frame the implications of his findings about civilian nonviolent strategies in conflict for warfighters, such as state militaries. But it does so by explicitly acknowledging the ethical tension involved—even when advocating for restraint—in engaging with actors that can possibly contribute to harm. As he writes, "Before considering the implications that civilian autonomy theory holds for counterinsurgency, it is important to address whether there is a place for them in a book that puts nonviolent civilian agency at the forefront. This is a delicate issue, especially because the goal of civilian autonomy is to deescalate war, and distilling implications for warfighters could possibly do the opposite ... This is not to say, however, that this text encourages counterinsurgency campaigns or any kind of belligerency."

change how actors behave in the real world, and not necessarily for the better. At the stage of sharing findings with policymakers, researchers' options to control how their findings are interpreted and used (or ignored) can quickly narrow. This may not matter if the policy process "box" is not really that black—if scholars have visibility into the policy process and know how their research is likely to be used. Researchers and policymakers may also be in agreement and have aligned interests and incentives over a policy issue. However, when there is uncertainty and the stakes are possibly high, there may be little a researcher can do to shape policy outcomes once they "give up their goods"—that is, once they share their insights with policymakers and, in some cases, lend their names to particular policy positions or institutions. In such situations, perhaps the only recourse is to create a "hold-up problem" and "hold-up" or decline to share one's research.

The "hold-up" problem is a common feature of contract and bargaining theory in economics. It represents a form of an incomplete contract that can lead to suboptimal "underinvestment," or underprovision of production.[7] In the case of responsible policy engagement, when there are concerns that policymakers are not adhering to their advice, scholars may precisely seek to ensure an "underinvestment" or underprovision of the implementation of bad policy. If worst comes to worst, holding-up can represent a rare source of power in a context in which mere academics are selling their wares. When there are concerns of misuse of their insights, they can choose to withhold their expertise and advice.

The risk of going public with information and deciding to "hold-up" is illustrated in a quotation by journalist Mark Doyle from the BBC World Service from an interview he gave for the documentary *Ghosts of Rwanda* (PBS 2004) on the Rwandan Genocide. Recalling the actions of an especially heroic UN peacekeeper soldier who worked to rescue hundreds of Tutsi civilians, he made the conscious decision *not* to contemporaneously report on the peacekeepers' actions for fear that the information could be appropriated by malign actors, endangering the soldier and bringing his rescues to an end. Doyle says,

> I knew what [Peacekeeper] Mbaye Diagne was doing. I had a very, very strong suspicion—put it that way—of what he was doing. And had I investigated, I could have found out, but I didn't want to find out. I didn't want to say, 'There is a

[7] In the classic description, a firm that provides inputs to another product-producing firm can withhold supply or raise prices, encumbering the downstream firm's production (e.g., Williamson 1971; Rogerson 1992). Economists and business managers usually seek to solve or reduce hold-up problems so that production remains efficient and profit is maximized, either by writing more explicit contracts or by bringing production of the primary goods in-house (vertical integration).

> Senegalese officer saving people in this town.' You can imagine what the impact of that would have been. He would have been killed.[8]

Along these lines, as told by Caballero in his account, he held-up the Colombian presidency's efforts to misuse his analysis by resigning from his job.

In addition to a complete "hold-up" of information, another option could be to engage in a selective refusal to provide policy advice: a "partial hold-up." This could take the form of giving only *some* information related to a policy issue. Or, it could mean providing only *vague*, fuzzy details related to the policymakers' interests. This is what Lyall reports he had to do when interacting with US military officials about airstrikes and targets in Afghanistan, deciding to only provide them with general models and avoiding analysis of real-time data. He declined to provide a more specific and timely model, one that could have been used to enhance targeting capabilities. In that way, he worked to avoid playing any direct role in potentially causing unknown harm to people on the ground.

Academics should realize they have agency in their hold-up power, and that they can use it when needed. They are not obliged to share all they know or remain in harmful policy engagements. Sometimes it is a mistake to do so (after all, there are reasons why the academia–policy "gap" exists). However, scholars should be aware that it may come at the cost of losing influence by not being involved in the process or being unable to steer decision-making to better outcomes. Hold-up tactics may therefore be suboptimal from a practical point of view.[9] It could be better for researchers and policymakers to build trust and reach prior understandings through clear communication of their aims and concerns and through repeated interactions (i.e., writing a more "complete contract"). Researchers can initiate and test a relationship by first trickling information to decision-makers, waiting to see how they behave, how their information may be used, and whether they receive any feedback about the "black box" of the policy process.

Reaffirming Transparency Norms While Challenging Other Norms

Advanced training in the social sciences often urges academics to guard against overgeneralization or other forms of misleading presentation of

[8] Diange was eventually tragically killed by shrapnel from an explosion. Subsequent news reporting brought the peacekeepers' brave actions to light.

[9] Researchers should also consider whether holding-up information could bleed into a kind of prosocial lying to the extent that omitting research findings or withholding their participation could be construed as only telling a partial truth about an important policy issue.

findings. It encourages practices such as being explicit about the limits of studies and findings, the consideration of alternative explanations, and rigorous methods that help establish causal inference or constitutive forces. Within the positivist tradition at least, striving for objectivity should position researchers to embrace whichever conclusions may arise from the evidence.

Unfortunately, when it comes to the policy sphere, academics may sometimes abandon circumspection and efforts to minimize uncertainty, and instead embrace stridency and sensationalism. This is perhaps understandable from the perspective of being heard and gaining influence. Yet, the case studies by Epstein (Chapter 7) and DeMartino (Chapter 6) highlight the ways in which the neglect of some disciplinary norms can undermine the quality of policy debates and sow doubts among the public about the integrity of institutions and expert advice. These scholars find that a new set of disciplinary norms has emerged in relation to policy engagement that is antithetical to the academic enterprise. These more destructive norms include prosocial lying, the overstatement of findings, and neglect of counterfactual reasoning and alternative explanations. Rather than truly striving for objectivity, scholars may either intentionally or unintentionally conceal political commitments within their theoretical and methodological choices.

Remedies for overgeneralization and prosocial lying begin with stronger cultivation of critical consciousness regarding policy engagement. In the economics profession, DeMartino chronicled the extent to which policymakers and academics alike were so secure in their supposedly superior knowledge relative to the general public that they had few qualms about shaping mass behavior. This was true even if it came at the expense of an open discussion of the acute distributional consequences of certain policies (i.e., liberalized trade) or to the detriment of citizens' ability to receive accurate information and make decisions with personal autonomy (as in the case of the 2008 financial crisis). In the realm of international security, Epstein also found that even highly regarded academic researchers were prone to overstatement in the debate over NATO enlargement—in part because they had never critically examined their political priors embedded within their analyses. With the passage of time and the evolution of world politics, some of their earlier statements appear highly irresponsible in retrospect. It is completely acceptable for researchers to sometimes "get it wrong" given the uncertainties of research and the complexities of international politics. It is less acceptable to make forceful claims about things that are simply unknowable while pretending otherwise.

Formal disciplinary norms should be reinforced and taught as part of Responsible Engagement professional trainings. One way to instill these norms is to spell out the consequences of when they are not followed, such as by studying example cases and experiences (e.g., Epstein [Chapter 7], DeMartino [Chapter 6]). Researchers can also learn from the best practices of existing bridging organizations. The Union of Concerned Scientists is an example of an organization whose mission is to counter prosocial lying—or any kind of deviant use of science, for that matter. They see "exposing" faulty science and policy implications as part of their mission. As their website states, "When special interests release bogus studies, corporations sow doubt about science, or political operatives censor, manipulate, or undermine government science, UCS is there to set the record straight with the public, the news media, and our elected officials."[10]

Adopting Inclusive Engagement Practices

Increasingly, social scientists are striving to include research "subjects," participants, and collaborators in the research process through such practices as co-designing and participatory action research. There are also continued efforts to include and credit scholars of diverse backgrounds (but there is certainly more road to travel). Nevertheless, the spokespeople—the heralds that often have the privilege of bringing the boons of socially beneficial research findings to the public sphere and to decision-makers—are often still the scholars with advanced degrees. Societal partners are not regularly involved in policy engagements, yet they can contribute unique perspectives and credibility. But there are potential challenges and costs to pursuing this inclusion. Kaplan and the partner organizations involved in the briefings on community-based atrocities prevention might have done more to reflect on the power balance between him and the local partners in their presentations. They could have also more explicitly anticipated and prepared for the costs and burdens imposed on the community-leader participant Cristina Serna by joining in the experience.

Responsible scholars must ask whether community, NGO, or other partners are at least being represented and credited in research briefings. A step further, they should also work to ensure these partners have seats at the policy table. To do this fairly and equitably, scholars must first reflect on their own positionalities: how they are viewed by others, possible power asymmetries, and what they bring. This contributes to a better understanding of relationships with engagement partners. Only by first looking within can

[10] See https://legacy.ucsusa.org/your-impact/.

balanced relationships be formed with partners to optimize cooperation in policy advising.

Recommendations for Policymakers

We are primarily a group of scholars, and our focus is primarily on researchers' responsibility. But policymakers are the key and obvious interlocutors of scholarly engagement. We conceive of "policymakers" in broad terms, to include national governments, local governments, communities, NGOs, businesses, or any other entity or individual involved in making or implementing collective policy decisions based on evidence. As former policymaker Erdberg Steadman asserts, some of the burden to achieve sound research-informed policy also lies with policymakers. There are several proactive steps policymakers can take to handle research findings more responsibly—as well as several actions they can avoid.

A first recommendation is for policymakers to become educated about scholars' motivations and pressures, as well as how scholars are coming to incorporate responsible practice into their research projects. They should be aware that the coming generation of scholars is going to enter into policy relationships with ethics at the forefront of their minds. They are more likely to ask to be informed about how their research may be used or that inclusive practices be followed to adequately involve research partners. They are also more likely to have ethical redlines and pull the metaphorical "ripcord" if interactions diverge from their expectations.

Policymakers should also be mindful about their own circumstances. They should recognize that they may face political and social pressures, both internally, within their organizations, and externally, to shun certain research findings or steer toward particular policy decisions—ones that may be suboptimal from a researcher's perspective (i.e., they may come to see some scholarly findings as "unpopular"). While policymakers, depending on their positions, cannot always avoid political pressure concerning research findings, they can at least anticipate possible pressures. In such cases they may be able to adapt their approaches or inform researchers upfront about the politics of the situation. They can signal their constraints in terms of accepting or acting on certain kinds of research results.

Policymakers and policy-implementing organizations can also articulate their own "responsibility" rules and codes, and perhaps even their own redlines. Some humanitarian organizations have published their own Codes of Conduct and guidelines on Accountability for Affected People (AAP) to protect their various stakeholders, especially after several scandals broke in recent years that led to harm of would-be beneficiaries.[11] The US Director of National Intelligence also developed a code for intelligence community (IC) organizations: "Principles of Professional Ethics for the IC [intelligence community]" (Clapper 2015). Some of the principles echo the responsible engagement guidance for scholars. For instance, the second principle hearkens back to the admonitions to avoid prosocial lying and overgeneralization: "We seek the truth; speak truth to power; and obtain, analyze, and provide intelligence objectively."[12]

Lastly, policymakers should strive to increase their awareness of their own possible biases related to responsible procurement and consumption of evidence generated by scholars and even research participants. Are they only cherry-picking academic experts whose views coincide with their priors or their desired policy outcomes? Or are they working to be inclusive of perspectives and backgrounds? Are they being transparent and open with researchers about the nature of the policy process and the possible end uses of academic findings? Are they representing views and findings accurately? Are they inclined to give feedback to academics, when feasible, about how their information was eventually used and how engagements panned out? Are they aware of their own defaults and values when it comes to implementing certain types of policy, especially when facing uncertainty or competing claims?

Recommendations for Donors

Donors can play critical roles in enabling, supporting, and shaping research that otherwise might not be conducted for lack of resources. They can also encourage policy relevance and help with publicizing results and connecting scholars with policymakers to share insights. However, funders of research may not necessarily emphasize researcher responsibilities and, in some situations, they can also put pressure on scholars in ways that could overlook ethical concerns. When grants include strict timelines to complete complex research projects, scholars may not have the time or resources to plan for or carry out responsible engagement practices. Similarly, donors may prioritize displaying the usefulness of their spending. They could place pressures on scholars to generate "impact" and share their findings, even

[11] Perhaps most notably, Oxfam staffers were involved in sexual misconduct when providing relief after an earthquake in Haiti (Rosenberg 2018).

[12] There is no lost hint of irony since the IC is often considered part of the power structure. Of course, whether and how much IC agencies or organizations in other domains adhere to such guidance is not necessarily visible to external observers or the general public (e.g., Froomkin 2014).

when the nature of the findings or policy interlocutor could raise ethical concerns.

Scholars who contemplate engaging in policy advising often wonder if it is possible to remain independent while receiving external support, whether the funding comes from a foundation, government, or another source. Some sponsored research can raise concerns if the research participants or collaborators of a study perceive a funding source as being politically or ethically suspect. We do not provide firm guidelines on what sources of research support may be more or less ethically upstanding, but we encourage the pre-registering of "ethical redlines" for funding and reaching out to mentors and peers for advice and accountability. In addition, opening lines of communication with sponsors and managing expectations can help prevent ruptures if research produces unpopular findings.

Donors can also play a positive role by supporting responsibly engaged research. They can explicitly provide resources for researchers—including their research teams and participants—to have the time and capacity to conduct responsible reflection and engagement at all stages of a project: from prior to the start of research all the way through to project completion and beyond. The Transforming Evidence network[13] of scholars and practitioners has helped to open the "black box" of donor strategies. They have also generated suggestions for how donors can be better supporters of relational practices to improve the policy uses of research (Tseng et al. 2022). Some advice includes promoting "relational" approaches to policy-engaged research that are more partnership-oriented. This parallels Kaplan's case (Chapter 5, this volume), though he more explicitly details cooperation during the briefing of results and places greater emphasis on responsibility and the risks of engagement.

A Research Agenda on Responsible Engagement

As an emerging topic, social scientists—ourselves included—are still coming to grips with what it means to "engage responsibly" and working to identify the many issues and forms of such engagements. The topic is therefore ripe for ongoing research to theorize different approaches to responsible engagement and collect empirical evidence on different models of practice. We identify several possible productive areas.

[13] See https://transforming-evidence.org/.

A first question meriting further inquiry is perhaps the most fundamental to this endeavor: What can we expect of responsible engagement practices in terms of effects and outcomes, and what does "success" mean in this context? Are there particular ethical approaches that are most useful for selecting criteria to assess successful responsible engagement? For instance, utilitarianism might equate responsibility with the maximization of (society-wide) benefits, while Kantian ethics might argue that scholars will contribute to the best outcomes when they advocate for decision-making processes that are equitable and autonomy-respecting. Rawls's (1971) justice-as-fairness approach (the original position and veil of ignorance) might call upon scholars to anticipate the differential effects of their advising and ensure they are elevating the concerns of the most vulnerable stakeholders. Certainly, there are many other relevant lenses and criteria for assessing success, as well.

These questions lead to several follow-on questions: How can different responsibility approaches, training, and efforts be measured and evaluated in terms of their effects on engagements? Does responsibility mean to simply "do no harm" or "minimize harm" against some counterfactual scenario of not following responsibility principles? Does it mean that certain responsible behaviors or practices are observable among scholars, such as decisions to halt partnerships or pursue reflections and dialogues?

Apart from the conceptual questions, more concrete studies can be conducted of policy engagement dilemmas and solutions, to include those herein but looking beyond them, as well. The chapter-cases of this volume are initial forays into their topics. The cases also each involve particular types of policy actors and are centered on single issue-areas, whether economics, national security, atrocities, corporate responsibility, peacebuilding, or development. Additional case studies could explore the generalizability of the dilemmas by checking if the same dilemmas play out similarly or not across different issue domains and policy actors. There are also no doubt many other yet-to-be-identified issues and dilemmas. One productive path could be for scholars to reflect on and generate additional case studies of their engagement experiences, perhaps expanding upon—though also rounding out—some of the principal–agent-style challenges we identify.

Public opinion studies are also a productive avenue. We have referenced some initial studies that examine researchers' attitudes toward engagement (e.g., Hendrix et al. 2023), but there is much more to learn, and attitudes and perspectives on practices are ever-evolving. There may be opportunities to generate additional insights about the views of policymakers, donors, and other engagement partners on issues of responsibility. The views of the general public may also be worth measuring since responsible practices may shape how citizens assess scientific credibility and public debates.

There are also promising opportunities for gathering feedback from policymakers and partners on the "black box" for evaluating the effects of responsible engagement experiences. While policymakers may not always offer candor, under less sensitive circumstances they may be willing to share their views on how findings and scholarly responsibility can influence policy decisions and implementation (as suggested by Kaplan [Chapter 5] and Erdberg Steadman [Chapter 9]; see, for example, the policymaker survey by Avey and Desch 2014). There is certainly also much to learn from existing transdisciplinary organizations that have acted as the consciences for policy engagements over decades, such as the Union of Concerned Scientists and the Bulletin of the Atomic Scientists.

The Future of Responsible Engagement

In many respects, the future of responsible engagement is bright. We are now more aware of possible engagement dilemmas than ever before. Survey data indicates that many scholars are concerned about engaging responsibly and looking to do so. And policymakers are becoming increasingly educated about research approaches and appreciate diverse sources of knowledge (Avey et al. 2022). We can all become more self-aware and improve our ethical practice. We call on researchers across the social sciences—and even across the sciences in general—to be proactive in discussing and implementing responsible engagement practices.

In encouraging responsible engagement, we are not writing to wag our fingers, cast blame or shame, or deter productive policy engagement. We hope to enhance engagements, protect scholars, help get policymakers the advice they need, and generate better outcomes all around. But, to broadly improve responsible engagement requires action at multiple levels. Change is important at the individual level, as each scholar can strive to be more reflective and perhaps more cautious in terms of how they share their research. But it is also important to take action together, as a scholarly community. Only then can we transform the academic culture toward engagement and shift norms in the field. And only then can we also shape the expectations of practitioner-consumers of research toward a more ethical practice.

These are not the last words on responsible engagement. Rather, they are some of the first. There are no doubt more responsible engagement dilemmas, models, and solutions to be discovered. Each of us has stories and experiences with valuable lessons. Write them down, reflect on them, share them, and, if appropriate, publish them.

References

Abrahms, Max (2018) *Rules for Rebels: The Science of Victory in Militant History*. Oxford: Oxford University Press.

Acemoglu, Daron, and James A Robinson (2012) *Why Nations Fail: The Origins of Power, Prosperity and Poverty*. New York: Crown.

Akoto, William (2020) "The 'Dual-Use' Dilemma in Policy Engagement." *Duck of Minerva*, August 19, https://www.duckofminerva.com/2020/08/the-dual-use-dilemma-in-policy-engagement.html.

Avey, Paul C. and Michael C. Desch (2014) "What Do Policymakers Want from Us? Results of a Survey of Current and Former Senior National Security Decision Makers." *International Studies Quarterly* 58(2): 227–246.

Avey, Paul C., Michael C. Desch, Eric Parajon, Susan Peterson, Ryan Powers, and Michael J. Tierney (2022) "Does Social Science Inform Foreign Policy? Evidence from a Survey of US National Security, Trade, and Development Officials." *International Studies Quarterly* 66(1).

Barma, Naazneen H., and James Goldgeier (2022) "How Not to Bridge the Gap in International Relations." *International Affairs* 98 (5): 1763–1781.

Barron, Patrick (2019) *When Violence Works: Postconflict Violence and Peace in Indonesia*. Ithaca: Cornell University Press.

Belew, Kathleen and Ramón A. Gutiérrez, eds. (2021) *A Field Guide to White Supremacy*. Oakland: University of California Press.

Bueno de Mesquita, Bruce and Alastair Smith (2011) *The Dictator's Handbook: Why Bad Behavior is Almost Always Good Politics*. New York: PublicAffairs.

Carpenter, Charli, Alexander H. Montgomery, and Alexandria Nylen (2020) "Breaking Bad? How Survey Experiments Prime Americans for War Crimes." *Perspectives on Politics* 19(3): 912–924.

Clapper, James R. (2015) "Remarks as Delivered by The Honorable James R. Clapper Director of National Intelligence: The Ethos and Profession of Intelligence." George Washington University, Washington DC, October 27. https://www.dni.gov/index.php/newsroom/speeches-interviews/speeches-interviews-2015/item/1278-remarks-as-delivered-by-dni-clapper-at-the-ethos-and-profession-of-intelligence.

Desch, Michael C. (2019) *Cult of the Irrelevant: The Waning Influence of Social Science on National Security*. Princeton: Princeton University Press.

Dukalskis, Alexander (2021) *Making the World Safe for Dictatorship*. Oxford: Oxford University Press.

Froomkin, Dan (2014) "Clapper Denies Lying, Announces New Ethics Policy." *The Intercept*, September 18. https://theintercept.com/2014/09/18/clapper-denies-lying-announces-new-ethics-policy/.

Gartenstein-Ross, Daveed and Varsha Koduvayur (2019) "How to Win Friends and Wage Jihad." *Foreign Affairs* (July/August), https://www.foreignaffairs.com/articles/middle-east/2019-07-01/how-win-friends-and-wage-jihad.

Hendrix, Cullen S., Julia Macdonald, Ryan Powers, Susan Peterson and Michael J. Tierney (2023) "The Cult of the Relevant: International Relations Scholars and Policy Engagement Beyond the Ivory Tower." *Perspectives on Politics* 21(4): 1270–1282.

Fortin, Jacey (2014) "At Quiet Rebel Base, Plotting an Assault on South Sudan's Oil Fields." *The New York Times*, April 3.

Jensen, Michael C. and William H. Meckling (1976) "Theory of the Firm: Managerial Behavior, Agency Costs and Ownership Structure." *Journal of Financial Economics* 3(4): 305–360.

Kaplan, Oliver (2017) *Resisting War: How Communities Protect Themselves*. Cambridge: Cambridge University Press.

Luttwak, Edward (1968) *Coup d'état: A Practical Handbook*. New York: Penguin.

Luttwak, Edward (1999) "Give War a Chance." *Foreign Affairs* 78(July/August): 36–44.

Metz, Cade (2023) "'The Godfather of AI' Leaves Google and Warns of Danger Ahead." *The New York Times*, May 1.

Mitnick, Barry M. (1975) "The Theory of Agency: The Policing 'Paradox' and Regulatory Behavior." *Public Choice* 24: 27–42.

Neustadt, Richard E. (1960) *Presidential Power: The Politics of Leadership*. New York: John Wiley and Sons, Inc.

Nincic, Miroslav and Joseph Lepgold, eds. (2000) *Being Useful: Policy Relevance and International Relations Theory*. Ann Arbor: University of Michigan Press.

Office of the Director of National Intelligence (2015) "Bin Laden's Bookshelf." https://www.dni.gov/index.php/features/bin-laden-s-bookshelf?start=5.

PBS (2004) *Ghosts of Rwanda*. Frontline, April 1. https://www.pbs.org/wgbh/pages/frontline/shows/ghosts/etc/script.html.

Pielke, Roger A. (2007) *The Honest Broker: Making Sense of Science in Policy and Politics*. Cambridge/New York: Cambridge University Press.

Rawls, John (1971) *A Theory of Justice*. Cambridge: Harvard University Press/Belknap Press.

Rogerson, W. P. (1992) "Contractual Solutions to the Hold-Up Problem." *The Review of Economic Studies* 4(59): 777–793.

Rosenberg, Eli (2018) "Oxfam Admits Staffers Engaged in Sexual Misconduct During Earthquake Relief Effort in Haiti." *The Washington Post*, February 9.

Ross, Stephen A. (1973) "The Economic Theory of Agency: The Principal's Problem." The *American Economic Review* (63)2: 134–139.

Sanford, Victoria, and Asale Angel-Ajani, eds. (2006) *Engaged Observer: Anthropology, Advocacy, and Activism.* New Brunswick: Rutgers University Press.

Semana (2008) "Alias 'HH' entrega testamento de Carlos Castaño." July 8.

Stokes, Emily (2009) "Lunch with the FT: Rory Stewart." *The Financial Times*, July 31.

Tseng, Vivian, Angela Bednarek, and Kristy Faccer (2022) "How Can Funders Promote the Use of Research? Three Converging Views on Relational Research." *Humanities and Social Sciences Communications* 9: 219.

Walt, Stephen M. (2005) "The Relationship Between Theory and Policy in International Relations." *Annual Review of Political Science* 8: 23–28.

Williamson, Oliver E. (1971) "The Vertical Integration of Production: Market Failure Considerations." *The American Economic Review* 61(2): 112–123.

Index

For the benefit of digital users, indexed terms that span two pages (e.g., 52–53) may, on occasion, appear on only one of those pages.